Mining and Natural Hazard Vulnerability in the Philippines

T0249062

Mining and Natural Hazard Vulnerability in the Philippines

Digging to Development or Digging to Disaster?

William N. Holden and R. Daniel Jacobson

ANTHEM PRESS
LONDON · NEW YORK · DELHI

Anthem Press
An imprint of Wimbledon Publishing Company
www.anthempress.com

This edition first published in UK and USA 2013
by ANTHEM PRESS
75–76 Blackfriars Road, London SE1 8HA, UK
or PO Box 9779, London SW19 7ZG, UK
and
244 Madison Ave. #116, New York, NY 10016, USA

First published in hardback by Anthem Press in 2012

British Library Cataloguing-in-Publication Data
A catalogue record for this book is available from the British Library.

Library of Congress Cataloging-in-Publication Data
The Library of Congress has cataloged the hardcover edition as follows:
Holden, William N. (William Norman), 1962–
Mining and natural hazard vulnerability in the Philippines : digging
to development or digging to disaster? / William N. Holden and R.
Daniel Jacobson.
p. cm.
Includes bibliographical references and index.
ISBN-13: 978-0-85728-776-2 (hardback : alk. paper)
ISBN-10: 0-85728-776-1 (hardback : alk. paper)
1. Mineral industries–Environmental aspects–Philippines. 2. Mines
and mineral resources–Environmental aspects–Philippines. 3. Natural
disasters–Philippines. 4. Mine accidents–Philippines. 5.
Environmental risk assessment–Philippines. I. Jacobson, R. Daniel.
II. Title.
TD195.M5H65 2012
363.73'1–dc23
2011047725

ISBN-13: 978 1 78308 051 9 (Pbk)
ISBN-10: 1 78308 051 5 (Pbk)

This title is also available as an ebook.

In memory of Eliezer "Boy" Billanes, 1962–2009

Of all those expensive and uncertain projects, however, which bring bankruptcy upon the greater part of the people who engage in them, there is none perhaps more perfectly ruinous than the search after new silver and gold mines.

—*Adam Smith*

CONTENTS

ACKNOWLEDGMENTS

The authors of this book owe a tremendous debt of gratitude to a very large number of people to whom they would like to articulate their sincere appreciation. For those not listed here, any errors and omissions are completely our own. First, and foremost, the authors would like to thank the editorial staff of Anthem Press, particularly Robert Reddick, Janka Romero and Tej Sood, for their assistance. The authors would like to articulate their gratitude to Meriam Bravante, Andy Whitmore and Sister Susana Bravante O. P. for arranging some of the most important interviews upon which this book was based. The literature for this book, which was found by the authors in hardcopy, was photocopied by the University of Calgary's document delivery staff while Brenda Paschke tirelessly printed the literature found by the authors in electronic form. Puerto Pension, in Puerto Princesa City, provided an ideal atmosphere for conducting a review of this literature. The discussion of acid mine drainage was made possible with the assistance of David Stiller. Kirsten Moran assisted with a review of the engineering literature on tailings dam stability. Ann-Lise Norman provided assistance with the literature on climate change. Shawn Marshal provided invaluable assistance in discussing possible links between climate change and increases in tectonic activity. John Yackel provided assistance with the discussion of the El Niño Southern Oscillation. Tom Dawson is to be thanked for keeping the computers running and Davor Gugolj is to be thanked for his assistance in finding maps layers. Robin Poitras deserves special thanks for his expert cartographic skills and his invaluable assistance with the artwork in this book. Two anonymous reviewers provided some extremely helpful comments and suggestions that profoundly impacted the final version of this book and, accordingly, these reviewers are to be thanked for taking the time to review this book and provide feedback. Penultimately, this book was heavily based upon fieldwork interviews and, accordingly, the authors would like to thank all of those innumerable individuals who so graciously provided their time by being interviewed. Last, and by no means least, the authors would like to thank their respective families for their patience and support; Meriam deserves credit for her inexhaustible supply of patience and her role as a sounding board for many of the concepts in this book while Kathleen is to be thanked for her limitless understanding and cups of tea.

LIST OF TABLES AND FIGURES

Tables

Figures

LIST OF ACRONYMS

AMB	Alternative Mining Bill
AFP	Armed Forces of the Philippines
ASG	Abu Sayyaf Group
AO	administrative order
ARMM	Autonomous Region of Muslim Mindanao
ATM	Alyansa Tigil Mina (Alliance to Stop Mining)
BNPP	Bataan Nuclear Power Plant
BEC	Basic Ecclesial Community
CAFGU	Citizens Armed Forces Geographical Unit
CAR	Cordillera Administrative Region
CARP	Comprehensive Agrarian Reform Program
CBCP	Catholic Bishops Conference of the Philippines
CEC	Center for Environmental Concerns
CHR	Commission on Human Rights
CPA	Cordillera Peoples Alliance
CPLA	Cordillera People's Liberation Army
CPP	Communist Party of the Philippines
DAR	Department of Agrarian Reform
DENR	Department of Environment and Natural Resources
DILG	Department of Interior and Local Government

DOJ	Department of Justice
DTI	Department of Trade and Industry
ECC	Environmental Clearance Certificate
EIA	environmental impact assessment
EIS	environmental impact statement
ELAC	Environmental Legal Assistance Center
EMB	Environmental Management Bureau
ENSO	El Niño Southern Oscillation
FMR/DP	final mine rehabilitation/decommissioning plan
FTAA	Financial and Technical Assistance Agreement
GDP	gross domestic product
GIS	geographical information systems
GNI	gross national investment
HDI	Human Development Index
HMB	Hukbong Mapagpalaya ng Bayan (People's Liberation Army)
HUKBALAHAP	Hukbong Bayan Laban sa Hapon (People's Anti-Japanese Army)
JPICC AMRSP	Justice, Peace and Integrity of Creation Commission of the Association of Major Religious Superiors of the Philippines
MGB	Mines and Geosciences Bureau
MILF	Moro Islamic Liberation Front
MMTF	Mine Monitoring Trust Fund
MNLF	Moro National Liberation Front
MPSA	Mineral Production Sharing Agreement
NDFP	National Democratic Front of the Philippines
NEDA	National Economic Development Authority
NGO	nongovernmental organization

NIPAS	National Integrated Protected Areas System
NLSA	National Land Settlement Administration
NPA	New People's Army
OFW	overseas Filipino worker
PAWB	Protected Areas and Wildlife Bureau
PHIVOLCS	Philippine Institute of Volcanology and Seismology
pH	potential of hydrogen
PHP	Philippine peso
PKP	Partido Komunista ng Pilipinas
PNP	Philippine National Police
SCAA	Special CAFGU Active Auxilliary
SOCCSKSARGENDS	South Cotabato, Cotabato, Sultan Kudarat, Sarangani, General Santos City, and Davao del Sur
SOP	standard operating procedure
TVI	Toronto Ventures Incorporated
UN	United Nations
UNDP	United Nations Development Programme
USD	United States dollar
WMC	Western Mining Corporation

INTRODUCTION

Phenomenon under Study: Mining amid Natural Hazards

The last three weeks of November 2004 and first week of December 2004 were a tumultuous time for the Republic of the Philippines. From 14 November until 4 December 2004 the Philippines was visited by four serious weather events: Typhoon Yoyong, Tropical Storm Unding, Tropical Depression Violeta and Tropical Depression Winnie (Yumul et al. 2011). These weather events, and the resulting flash floods and landslides, inflicted havoc on the Philippines killing 1,068 people, injuring 1,163 and causing another 553 to go missing; collectively, these storms caused roughly PHP 7 billion (USD 125 million) worth of damage to crops and infrastructure and also threatened to disrupt the water supply of Metro Manila. When the final damage total was completed the overall damage of these storms was estimated at PHP 11.3 billion (USD 201 million).

On 1 December 2004, while these storms were ravaging the Philippines, the Philippine Supreme Court issued a historic ruling[1] and reversed a decision it had made on 27 January 2004. In its 27 January 2004 ruling, the Supreme Court had invalidated some crucial provisions of the Mining Act of 1995 which allowed foreign corporations to own 100 percent of a mine located in the Philippines. This decision had complicated the efforts of the Philippine government to use large-scale mining as a development strategy. With the 1 December 2004 ruling reversing the earlier invalidation of the Mining Act of 1995 the statute was, in effect, validated and the government felt it had cleared the final hurdle to embark upon an aggressive promotion of large-scale mining.

While the coincidence of these storms with the 1 December 2004 court case may appear to be unrelated, the simultaneity of these storms with a court case facilitating a mining-based development paradigm reveals a vitally important question: what are the difficulties inherent in attempting to pursue a mining-based development paradigm in a country beset by natural hazards?

---

1 *La Bugal-B'laan Tribal Association Inc et al. v. Ramos et al.*, G.R. No. 127882.

Mining is an activity with a substantial potential for environmental harm. Mining involves the use and creation of highly toxic substances and can have impacts upon both water quality and water quantity; mines, being limited by the location of ore deposits, cannot be moved once developed. The Philippines is a country subjected to numerous natural hazards (such as typhoons, earthquakes, tsunamis, volcanoes and El Niño–induced drought) and is inhabited by poor people engaged in subsistence activities who are highly vulnerable to any form of environmental degradation. Could these hazards interfere with mining, worsen the conditions of the poor and create disasters? Will a mining-based development paradigm generate development lifting the poor out of poverty or will it deprive them of their basic means of survival and generate disasters? As the title of this book asks, will this be an example of "digging to development" or will this be an example of "digging to disaster?"

Neoliberalism: A Controversial Paradigm

In recent years, with the ascendency of neoliberalism, many developing countries have changed their legal systems in order to encourage foreign direct investment by multinational corporations (McCarthy 2007). An integral component of this has been a liberalization of mining laws to encourage mining by multinational mining companies (Bridge 2004, 2007). As this has occurred, there has been a tendency in much academic literature automatically to condemn neoliberalism (in general) and mining (in particular) as a failed method of accelerating development (Ward and England 2007). Wisner et al. (2004, 54) are highly critical of this tendency to automatically condemn mining and have articulated a view that any proper examination of mining's efficacy requires a thorough examination of each developing country's "particular historical and spatial specificities." To answer a question properly, such as the one postulated here, there must be "thorough research that is locally and historically based" (Wisner et al. 2004, 54). This book addresses this question by examining the political economy of mining in the Philippines with political economy being defined as "the interrelationships between political and economic processes and their implications on resources and wealth" (Roque and Garcia 1993, 5). The book looks at the nature and structure of society in the Philippines wherein a small number of extremely affluent people have come to control society. Once this is established, attention turns to the environmental effects of mining and then juxtaposes these within the set of hazards and vulnerabilities present in the Philippines; what Mara and Vlad call "the cartography of risk areas" (2009, 968). It is ultimately argued that the power of those who control Philippine society is so immense – and the rewards they stand to receive from mining so substantial – that the poor and marginalized are gravely threatened by locating mining amid the hazards and that little, if any, benefit will accrue to them from mining.

The Disciplinary Location of this Book: Geography

The four traditions of geography

In a seminal 1964 article published in the *Journal of Geography*, William D. Pattison outlined four traditions of geography.[2] The first tradition is the spatial tradition concerned with displaying where things are located and is apparent in geography's proclivity to use maps. The second tradition is the area studies tradition, best displayed by geography's subfield of regional geography which sums up and regularizes knowledge "of the nature of places, their character and their differentiation" (Pattison 1964, 213). The third tradition is the human-environment tradition, "a balanced tracing out of interaction between [humans] and environment" (Pattison 1964, 214). Finally, the fourth tradition is the earth sciences tradition, long revealed in geography's subfield of physical geography – which is "the study of the earth, the waters of the earth, the atmosphere surrounding the earth and the associations between the earth and sun" (Pattison 1964, 215).

While this book – to varying degrees – involves aspects of all four traditions,[3] its main focus is on the relationship between humans and their environment; it clearly finds its home within the discipline of geography (in general)[4] and within the human-environment tradition (in particular) as

2 Sluyter et al. (2006, 594) referred to Pattison's 1964 article as a "classic" while Gauthier and Taaffe (2002, 523) wrote that Pattison's four traditions "still provide a reasonably good description of the subject matter treated by geographers since 1900."

3 The first tradition is represented in an extensive use of maps, the second tradition appears in the discussion of the nature and character of the Philippines, and the fourth tradition clearly manifests itself in the discussion of the natural hazards present in the Philippines.

4 Indeed a highly geographic methodological approach was employed to capture the spatial specificities of mining in the Philippine context: the use of geographical information systems (GIS). Information on the location of mines was obtained from documents published by the Mines and Geosciences Bureau, of the Philippine government's Department of Environment and Natural Resources, and this information was then overlaid upon maps of natural hazard locations, the data for which was obtained from the website of the Manila Observatory. This approach was then combined with a series of interviews conducted with key informants (government officials, environmental activists, human rights activists and members of communities adjacent to mining projects) to provide insights into the particular social, historical and geographical characteristics of the Philippines in answering the question under examination.

"one of human geography's core areas [is the relationship] between people and their environment" (Bednarz 2006, 243). As Gray and Moseley (2005, 10) wrote, "geography, with its long-standing human-environment tradition, has produced a prodigious amount of scholarship regarding the factors that influence resource management and human-environmental interactions" (2005, 10). Perhaps the most famous example of this tradition was the book *Man's Role in Changing the Face of the Earth* (Thomas et al. 1956). This encyclopedic collection of essays examined the relationships between humans and their environment and still stands as a landmark of geographical literature.

Mining: A fruitful topic of geographical research

A subset of geography's long tradition of examining human-environment relations is an examination of mining (Bakker and Bridge 2006; Coban 2004). *Man's Role in Changing the Face of the Earth* contains a chapter entitled "Man's Selective Attack on Ores and Minerals," which discusses mining's environmental effects (McLaughlin 1956). The inclusion of this chapter in a landmark compilation demonstrates that mining is very much a subject of concern to geographers. Every aspect of mining, from the location of ore deposits through to the sale of processed minerals in global markets, contains a geographical dimension and this reveals one of the classic themes in geography: the relationship between the local and the global (Swyngedouw 2004). Just as mining has constituted an essential component of human-environment relations, so has the study of conflicts about mining as these are often conflicts about "two different understandings of human-nature relationships" (Coban 2004, 454). Conflicts over resources "form the empirical heart of some of the most theoretically informed research by geographers in recent years: most often, however, this work flies under the flag of political ecology" (Bakker and Bridge 2006, 7).

Political ecology: A geography-based research field

Political ecology is a research field frequently – but not exclusively[5] – found in geography that explores the political dimensions of human-environment interaction (Bryant and Bailey 1997). The term political ecology was first used by Wolf (1972) to demonstrate how power relations among humans mediate human-environment relations. Political ecology differs from traditional cultural ecology that focuses, in an apolitical manner, on the problems of adaptation to environmental change without examining structures of social

---

5 Many political ecologists are anthropologists (Bryant and Bailey 1997).

inequality occasioning environmental change (Biersack 2006). In contrast, political ecology shows how environmental analysis and policy can be reframed towards addressing the problems of the socially vulnerable (Forsyth 2008). Throughout this book the interplay of societal power relations and environmental change appears frequently, both in the sense of how power relations within society give rise to environmental change and in the more subtle sense of how environmental change impacts power relations within society rendering some weaker and others stronger.

The Demarcation of an Important Caveat

At the outset of this book the demarcation of an important caveat is in order: when this book discusses "mining" it is referring to large-scale mining carried out by corporations. In the Philippines it is the express policy of the government to encourage large-scale mining by multinational corporations and increasingly, in today's world much economic activity is being conducted by such corporations; entities who have no real physical home and who operate worldwide responding to no single authority, accounting only to their shareholders and focusing only on their profitability. Accordingly, artisanal – or small-scale – mining is beyond the scope of this book. "Artisanal miners" are defined by McMahon et al. (1999, 3) as "very small miners with little or no mechanization." These miners use simple tools such as sluice boxes, gravel pumps, bowl mills and air compressors (Tujan and Guzman 2002). The Philippines has a long tradition of small-scale mining, such as the Igorot pocket miners of northern Luzon, dating back earlier than the recorded history of the country (Broad and Cavanagh 1993). While small-scale mining does indeed pose challenges to the environment, its environmental effects are of a different degree and magnitude from those of large-scale mining. With this thought in mind, this book refrains from focusing on small-scale mining and directs its attention towards large-scale mining conducted by corporations.

The Outline of the Book

The book is laid out in the following manner: first, the reader is introduced to the Philippines, a land of extreme poverty dominated by a powerful oligarchy yet endowed with substantial mineral wealth. Then the efforts of the government to accelerate the economic development of the country by encouraging the extraction of minerals are discussed. This is followed by an explanation of the environmental effects inherent in modern mining; these effects are anything but minor and short lived. The natural hazards present in the Philippines are examined and the interaction of these hazards with mining is discussed. Two highly technocratic solutions to the problems posed by mining amid natural hazards (the subjection of mining projects to an environmental impact assessment process and the mining industry's commitment to best practices in environmental management) are presented and critically examined, particularly in view of how modern humans live in a risk society wherein the greatest dangers to humans come not from what nature may do to us, but from what we may do to ourselves. Finally, the book concludes with a discussion of how, given the risks posed to mining by the natural hazards present in the Philippines, mining may be a flawed development paradigm and asks whether another world is possible to that offered by corporate driven neoliberal globalization as embodied by mining.

This book is primarily intended for readers interested in the political economy of mining in the Philippines, but it may also be of interest to those interested in mining in general and those interested in natural hazards. Accordingly, given a possible wide variety of readership, this book engages literature spanning several disciplines including: anthropology, development studies, economics, engineering, environmental science, geography, geology, history and political science. With several disciplines being engaged simultaneously, the authors have not assumed a substantial degree of specialization in any one discipline and this may lead to some sections seeming redundant. It is the authors' hope that this ensures that the book is informative to all.

Chapter One

MINING IN THE PHILIPPINES

The Philippines: A Developing Country in Southeast Asia

An introduction to the archipelago

The Republic of the Philippines is an archipelago of approximately 7,100 islands in Southeast Asia (Figure 1.1) located on the western side of the Pacific Ocean between latitude 20 degrees north and latitude 6 degrees south. The archipelago lies east of Vietnam, south of Taiwan and northeast of Borneo and consists of four distinct regions: the northern island of Luzon, the southern island of Mindanao, the long finger-like western island of Palawan and the central Visayan Islands. It has a land area of 300,000 square kilometers with approximately 65 percent of this being taken up by Luzon and Mindanao while only 500 of the 7,100 islands exceed one square kilometer in area (IBON 2002c). Located on the "Pacific Ring of Fire" – a belt of volcanoes running along the Pacific coasts of the Americas and Asia – the islands of the Philippines are mountainous and are frequently described as a series of "half drowned mountains" (IBON 2002c, 8). The majority of the land area of the archipelago consists of land over 350 meters above sea level with flat land remaining concentrated in river valleys or in small discontinuous strips of coastal plains backed by steep mountains (Newson 1999). Indeed, the topography of the Philippines could be described as an outcrop of mountains emerging from the sea with a small fertile coastal strip and a series of narrow interior valleys (Bankoff 2003a).

In 2011, the Philippine population stood at approximately 100 million people (United States Census Bureau 2011). As of 2007, approximately 50 percent of the population was living on the island of Luzon, 25 percent of the population was living on the island of Mindanao, and 19 percent was living throughout the Visayan Islands (National Statistical Coordination Board 2011). Like many countries in Southeast Asia, the Philippines has a large and rapidly growing capital city to which migrants from countryside move, and Manila houses almost 11 percent of the national population (National Statistical Coordination Board 2011). With a 2010 population growth rate of 1.9 percent per year, the population of the archipelago can be expected to double by the year 2047 (United States Census Bureau 2011).

Figure 1.1. The Philippines

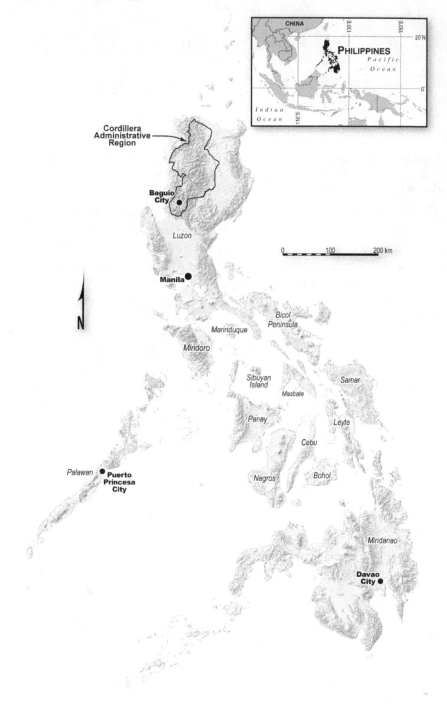

A developing country

A member of the so-called "developing world,"[1] the Philippines is classified as a "lower-middle-income country" by the World Bank (World Bank 2009). In 2009, its gross national income (GNI)[2] per capita was USD 1,620, a number below average relative to other East Asian countries (USD 2,182) and other lower middle-income countries (USD 1,905) (World Bank 2009). The Philippines, like many other developing countries, faces challenges such as low life expectancy, low levels of education and low levels of personal income. In this regard, the Human Development Index (HDI), a composite index taking into account life expectancy, education, and income can be a useful descriptive statistic. The HDI can range from a high of one down to a low of zero and, in 2010, the highest HDI score was Norway, at 0.938, while the lowest HDI score was Zimbabwe, at 0.14 (United Nations Development Programme 2011). The Philippines, with an HDI score of 0.638, ranked at 97 in the world, well behind number one ranked Norway and below the East Asia and Pacific average of 0.650 (United Nations Development Programme 2011). It must be stressed that this national HDI score of 0.638 does not reflect homogenous development throughout the archipelago and – as Figure 1.2 shows with its depiction of provincial HDI scores during 2006 – there are substantial variations in development from province to province, with Benguet having the highest HDI score of 0.787 and Sulu claiming the lowest HDI score of 0.326 (Human Development Network 2009).

A society dominated by an oligarchy

Inequality in the distribution of income is a prominent aspect of society in the archipelago. In the words of Tyner,

> The Philippines is a country of haves and have-nots. It is a country of the super-rich and the abject poor. It is a land where the richest 10 per cent of the population hold over 40 percent of the total income, while

---

1 The authors realize that the term "developing world" is a controversial term. However, so are the terms "Third World" (a relic of the Cold War with a "First World," the West, and a "Second World," the Soviet bloc) and "Global South" (the Philippines is, in fact, in the northern hemisphere while Australia, a highly developed country, is in the southern hemisphere). Given the controversy surrounding these descriptive terms, to describe the poorer countries of the world the term "developing world" has been selected, albeit with certain reservations.

2 "Gross national income" is defined as gross domestic product plus net receipts of primary income from abroad (World Bank 2009).

Figure 1.2. Human Development Index by province

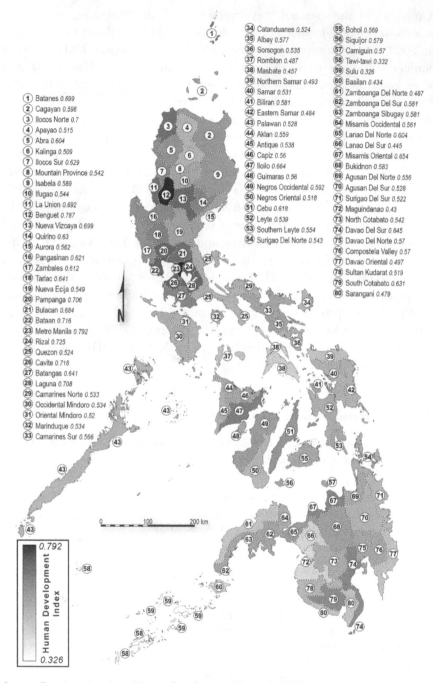

Source: Based on data from Human Development Network (2009).

the poorest account for less than two percent. Overall, nearly half of the population lives on less than US $2 a day. (2009, 2)

To quantify the distribution of income, economists use a measure of income inequality known as the Gini coefficient, which can range from a score of zero indicating perfect income equality, to a score of 100 indicating perfect income inequality. From 2002 to 2008, the Philippines had an average Gini coefficient of 44; a number giving it a higher degree of inequality than such neighbors as China (42), Indonesia (39), Malaysia (38), Thailand (42), and Vietnam (39) had over this same time period (World Bank 2010). The disparity in the distribution of income is a manifestation of the domination of the Philippines by a powerful oligarchy. There are many writers who regard the tremendous power of this oligarchy as being an obstacle to any amelioration of the conditions of the poor (Hawes 1987; Putzel 1992). In the words of Bello et al. (2009, 79), "The national policies of the government have always favored economic and political elites, thereby entrenching poverty and social and economic inequality." Attention now turns to an examination of the historical origins of this oligarchy.

Class structure in pre-Hispanic society

Prior to the arrival of the Spanish in the sixteenth century[3] the population of the islands lived in small coastal villages consisting of 100 to 500 inhabitants; these villages were called *barangays* after the Malay term *balangay* (which were boats used by the original Malay settlers of the archipelago) and their inhabitants relied upon fishing for sustenance (Constantino 1975). At the time of the arrival of the Spanish, the *barangays* were already experiencing a process of rudimentary social stratification and were ruled over by a *datu* – or chieftain – and a group of *maharlika* – or nobles – with the vast majority of people consisting of a group of people known as *alpin* who were often held in a form of debt-bondage (Tyner 2009). However, it must be emphasized that the barriers between these classes were not rigid ones and there was no concept of individual freehold property in land (Putzel 1992).

The Spanish colonial period, 1568–1896

It was with the coming of the Spanish that the entrenchment of a powerful oligarchy began to emerge as a dominant feature of Philippine society; "The

---

3 The Spanish originally became aware of the Philippines when Ferdinand Magellan claimed it for Spain in 1521 but it was not until the expedition of Miguel Lopez de Legazpi in 1565 that the Spanish set about colonizing the archipelago.

Spanish colonial regime left a legacy of plutocracy where a few hundred families control the country's resource system" (Roque and Garcia 1993, 104). One of the first things the Spanish did upon taking control of the archipelago was institute a land administration system known as the *encomienda*, where blocks of land were entrusted to the control of representatives of the crown; then, over time, these *encomiendas* began to be replaced by *haciendas* where *hacendados* were given ownership of the land complete with the right of inheritance and free disposition (Constantino 1975). Originally, most of the *haciendas* went to *conquistadores* but before long many religious orders within the Roman Catholic Church, such as the Augustinians, Dominicans, Jesuits and Recollects began to take control of the best agricultural lands, and by the end of the Spanish colonial period these friar lands totaled approximately 170,000 hectares (Roque and Garcia 1993). Enriched by its ownership of agricultural lands, the church became a powerful institution (*La Frailocracia*) in Spanish colonial society "which arrogated unto itself the political prerogatives and authority of the state" (Tan 2002, 17). Indeed, the church was so dominant during Spanish colonial times that Tan has referred to this period as "four centuries of brainwashing by catechetical methods" that "regimented the compliant population into devotion to various icons of Spanish religiosity" (2002, 17). "The priest used his role as father confessor to force his parishioners into debt and further dependence upon him" and only rarely "did clerics take a real stand to protect their parishioners rights" (Nadeau 2008, 29).

During the Spanish colonial period society began to segregate into three distinct groups (Putzel 1992). At the apex of the colonial social order were the *peninsulares*, Spaniards born in the Iberian Peninsula, who, along with the religious orders, occupied key positions in the state. Below these people were the *insulares*, Spaniards born in the islands, and the *datus* who commanded authority in the *barangays* by collecting tribute and organizing forced labor for the crown; collectively these people became known as the *principalia*, or "prominent ones." At the base of society were the *tao*, the vast bulk of the population of Malaysian ancestry referred to by the Spanish as either *indios* or *naturales*.

One group that began to assume an increasing degree of importance after the first 200 years of Spanish colonial rule was the Chinese (Hawes 1987; Putzel 1992). Originally, Spain attempted to keep the archipelago closed to most of the outside world and no Spaniards living in the islands were allowed to visit Asian ports. Although Chinese traders from Fukien and Kwantung were looked down upon by the Spanish and derogatorily referred to as "*sangleys*" or "merchants," they were attracted to Manila where they would trade silks and pottery for gold and silver transported by the Manila galleon from Mexico and, by the early eighteenth century, Chinese merchants began intermarrying with members of the *principalia* and in so doing created a *mestizo* land owning class. As the Spanish began to make Philippine agricultural crops such as indigo, sugar and abaca available for export, there was a rapid growth in agricultural exports from the archipelago. This growth in exports surged after the opening of the Suez Canal in 1869, which dramatically reduced travel time between the Philippines and Spain. With more exports there was more wealth for those owning land and the land owning class began to consolidate its power in society.

As the *mestizo* land-owning class became richer it also began to acquire more land through the use of the *pacto de retroventa* (Constantino 1975; Putzel 1992). *Mestizo* landowners would lend money to peasant small holders and these loans would be secured by the lender taking control of the borrower's land. If the borrower repaid the loan the land would revert back to the borrower but if the borrower defaulted on the loan, as was more often the case, the land became the property of the lender. Through the skillful use of the *pacto de retroventa* by *mestizo* landowners, thousands of small landowners became disposed of their lands and the rich and powerful acquired more land and were able to receive even more revenue from the production and sale of agricultural exports. At this time some of the powerful families (the Aquinos, Ayalas, Cojuangcos, Laurels, Lopezes, Sorianos, and Zamoras) that still dominate the archipelago today became well established. As Putzel (1992, 49) wrote, "Three hundred years of Spanish colonial rule had given rise to a landed oligarchy whose fortunes were largely tied to export agriculture."

The revolution of 1896

The Spanish colonial era was by no means a period of tranquility and it was marked by numerous peasant uprisings. On the island of Bohol, for example, the Spanish lost complete control from 1744 until 1829 when peasants established mountainous communities and defied the Spanish (Constantino 1975). As the twentieth century approached, however, the conditions faced by the peasantry increasingly began to worsen; as Linn (2000, 16) wrote, "By the 1890s much of the Philippines was in severe distress, plagued by social tension, disease, hunger, banditry and rebellion." The waning years of the nineteenth century also saw developing unrest among members of the upper classes as they began to agitate for change due to grievances generated by the Spanish domination of colonial society (Hawes 1987). Members of the *principalia* felt discriminated against when their sons were passed over by the Spanish church hierarchy for placement in the priesthood and when administrative positions in the colonial government were reserved for *insulares*; this only worsened as Spain lost its Latin American colonies and Spanish administrators from those former colonies migrated to the Philippines creating an even more oppressive and overstaffed bureaucracy (Hawes 1987). By the 1890s educated members of the upper classes, referred to as the *ilustrados* ("enlightened ones"), began agitating for liberal reforms in the colonial administration and even for representation in the Spanish courts (Hawes 1987).

When revolution against Spain finally broke out in 1896, it was a revolution of both the upper and lower classes. As Putzel (1992, 50) wrote, "The Revolution of 1896 represented a juncture between the animosity of the emerging Filipino *ilustrado* elite against the friar orders and Spanish dominance, and the aspirations of the peasantry for *kalayaan*, or freedom." The revolution was initially spearheaded by an organization of landless tenant farmers known as the Katipunan[4] under the leadership of Andres Bonifacio, a man from a lower-middle-class background (Constantino 1975). "The *Katipunan* began in the Tagalog region and was composed of the dispossessed lower urban class who could never hope to gain a higher education" (Nadeau 2008, 41). Before long, though, a wealthy landowner named Emilio Aguinaldo took control of the Katipunan and had Bonifacio killed (Constantino 1975). Under Aguinaldo's leadership, the revolution against Spain became a revolution against Spanish control of the archipelago and lost any pretense of being a revolution capable of remaking Philippine society (Silbey 2007). The revolution was initially led by urban artisans and workers, along with small landholders and tenants from the friar lands around Manila, but eventually its leadership passed into the hands of the upper classes (Hawes 1987). According to Danenberg et al. (2007, 294), the Philippine revolution was a national liberation struggle that ended up as a victory only for the national bourgeoisie; what had started as a "struggle for national liberation turned out as liberation for the newly formed bourgeoisie." "The 1896 Revolution led to the overthrow of Spanish rule and the curtailment of friar authority, but the land tenure system was left very much intact" (Putzel 1992, 51). "The elite," wrote Hawes, "was not interested in self-liquidation, and so its members did not seek radical transformations in Philippine society. They were politically progressive but economically conservative" (1987, 23). "For the vast majority of people, the transition from Spanish revolutionary government in 1896 meant little more than a continuation of rule by the local elite" (Linn 2000, 196).

The American colonial period, 1898–1946

On 25 April 1898, in a conflict of very vague origins, the United States and Spain went to war (Silbey 2007). On 1 May 1898 the United States Navy

---

4 "Katipunan" is short for "Kataastaasan Kagalang-galang na Katipunan nang mga Anak ng Bayan" (Highest and Most Venerated Association of the Children of the Nation).

under the command of Commodore George Dewey defeated the Spanish
fleet at the Battle of Manila Bay, and that June the United States Army sent
2,500 troops to the islands under the command of General Wesley Merritt to
assist Dewey in securing the archipelago (Linn 1989, 2000; Silbey 2007). By 10
December 1898, when the Treaty of Paris was signed giving the Philippines
to the United States in exchange for USD 20 million, the United States had
made the decision to retain the islands (Hawes 1987). The Filipinos, who had
declared independence from Spain on 12 June 1898, were unwilling to accept
one new colonial master as a replacement for another and, in February 1899,
the Philippine–American War broke out and lasted for almost three years
(Linn 1989, 2000; Silbey 2007). The conflict between the nascent Philippine
Republic and the Americans was by and large led by the same members of
the *principalia* who led the revolution against Spain and the Americans were
ultimately able to prevail in this conflict by convincing the elite that their
interests were better served by allowing the Americans to assume control of
the archipelago than by resisting them (Tan 2002). "The local elite who led the
resistance eventually found the cost of guerrilla war too high to bear: theirs
were the lands sequestered; theirs were the tax assessments from both sides
and theirs were the families and fortunes at risk" (Linn 2000, 197). In the
words of Putzel (1992, 51), "The local landed oligarchy that dominated the
short-lived Philippine Republic soon came to an understanding with the new
colonial authorities."

One thing the United States introduced – and this greatly assisted their
co-optation of the elite – was the (supposedly) rich tradition of American
electoral democracy (Coronel et al. 2007). Even before the Philippine–
American War was concluded, the Americans held local elections and in 1907
the first elections were held for the Philippine Assembly. In this election only
1.4 percent of the 7.6 million Filipinos were allowed to vote and these voters
consisted of men who held posts during the Spanish regime and who were
able to speak, read, and write either English or Spanish. These elections
ensured the continuity of power of the prerevolutionary elite and the election
of a legislative body composed of representatives from single member
districts further entrenched the power of the landholding elite. During the
American colonial period, "democracy" was installed but this was sharply
circumscribed by a requirement for property ownership eliminated only in
1935 (Holden 2009a). By the time the franchise was widely expanded in the
late colonial period "the dominance of the national oligarchy was so well-
entrenched that challenges from below faced monumental odds" (Hutchcroft
2008, 142).

Much has been written about how the American Army during the Philippine–
American War provided social services such as education and health care

(Hawes 1987; Linn 1989, 2000). According to Linn (2000, 83), "public order, efficient local government, education, public health and communications were the hallmarks of US Army civic reform." However, while the United States did provide social services for the population it did nothing to address the deep-seated problems inherent in Philippine society. "Humanitarian officers," wrote Linn (2000, 83), "sought pragmatic, sensible objectives that provided immediate benefits; but left untouched the deep social problems of the archipelago."

Something the Americans made no attempt to implement was land reform and there were two main reasons for this (Putzel 1992). First, they possessed neither the spirit nor the knowledge to implement land reform effectively because it had not been a socioeconomic problem in their national experience. Second, whilst the United States was aware of the problems occasioned by the dominance of the landed oligarchy, American officials decided that their twin interests of the geopolitical location of the Philippines and an assured supply of tropical commodities would be best served by maintaining the archipelago's existing property relations. Perhaps the closest the Americans came to implementing land reform was the redistribution of the friar lands that they purchased from the Vatican for approximately USD 7 million and then resold, ostensibly, to those share tenants who were already using them. However, once the friar lands were resold they ended up being purchased by the landed oligarchy who ultimately strengthened their position in society even more (Putzel 1992). As Karnow (1989, 198) wrote, "The Americans coddled the elite while disregarding the appalling plight of the peasants, thus perpetuating a feudal oligarchy that widened the gap between rich and poor."

The combined effects of introducing electoral democracy while refraining from implementing land reform was to transform the archipelago's economic elite into both an economic and political elite that continues to wield power today; the combined Spanish and American colonial period gave rise to an economic and political system in which a relatively small group of extremely affluent families enjoy substantial power in society (Hutchcroft 2008; Putzel 1992). From the colonial experience of the Philippines we see a similar linkage between the Spanish and American colonial administrations and the evolution of the archipelago's social structure (Hawes 1987). By the time of independence on 4 July 1946, the political and economic systems of the Philippines were firmly in the hands of a landowning, agricultural elite generated by Spanish and American colonial practices (Hawes 1987). "The Americans," wrote Agoncillo (1990, 444), placed at the helm of government "the class which learned from and inherited the Spanish colonial's technique of mass exploitation."

The Bell Parity Amendments

Just as the Philippines acquired independence from the United States in 1946, an event occurred ensuring the further entrenchment of the oligarchic elite in Philippine society: the Parity Amendments to the Philippine Constitution. When the Philippines achieved independence from the United States in 1946 the archipelago was reeling from the effects of World War II. In 1946, the United States offered to provide rehabilitation assistance to the Philippines if three conditions stipulated in the Philippine Trade Act (or Bell Trade Act) were accepted by the Philippines: first, the acceptance of a trade regime where American products would be given unimpeded access into Philippine markets while Philippine products would be allowed into the American market subject to quotas; second, the tying of the peso to the dollar; third, a granting of parity rights to American investors in the Philippines ensuring that they would be given the same rights as Filipino citizens (Constantino and Constantino 1978; Lopez 1992).

The provisions in the Bell Trade Act granting American investors parity rights were extremely controversial. The 1935 constitution, which created the Commonwealth of the Philippines (and was to guide the islands into sovereignty), was a highly nationalistic document (Lopez 1992). Accordingly, any granting of parity rights would have required a constitutional amendment and this could only be done by a three-fourths majority of both houses of the Philippine Congress (Constantino and Constantino 1978). In 1946, as the vote in Congress was pending, a group from Central Luzon consisting of members of the Hukbong Bayan Laban sa Hapon (People's Army against Japan, or Hukbalahap, the most effective resistance organization during the World War II Japanese occupation of the islands) and the Pambansang Kaisahan ng mga Magbubukid (National Peasants Union) won seats in the House of Representatives as the Democratic Alliance (Holden 2009a). The Democratic Alliance sought to address problems such as poverty, agrarian unrest, and the development of a weak economy dependent on the United States (Holden 2009a). Needless to say, the Democratic Alliance was opposed to amending the constitution to allow parity rights for American investors (Constantino and Constantino 1978). In 1946 just before the vote was to take place, the members of the Democratic Alliance were prevented from taking their seats "pending investigation of charges of alleged fraud and terrorism in their election" (Constantino and Constantino 1978, 203). With the Democratic Alliance excluded from Congress the parity amendments passed and this had three lasting implications for the Philippines. First, even though quotas would be imposed on the importation of Philippine agricultural products into the United States, the producers of these products – the rural agricultural elite – supported this because they viewed it as an assurance that they would have access to the American market (Hawes 1987). Second, this gave American corporations a dominant position in the islands; as Rodriguez (2010, 10) wrote, "With its preservation of US economic and political interests, the newly 'independent' Philippine state was, in fact, a neocolonial one." Third, the barring of the Democratic Alliance from Congress gave rise (in the short run) to the Hukbong Mapagpalaya ng Bayan (Army of National Liberation or HMB) uprising and (in the long run) discouraged the left from attempting to use electoral politics as a vehicle for social change (Holden 2009a). As Coronel et al. (2007, 255) wrote: "For years, leftist groups would revisit this experience each time they entertained thoughts of engaging in parliamentary struggle." The parity amendments controversy was clearly the final inertial force from the combined Spanish and American colonial period ensuring a path dependence of the oligarchic elite, what McCoy (2009, 533) would call the polarization "between a wealthy oligarchy and an impoverished mass."

A poorly performing economy

Throughout the 1950s and 1960s, the archipelago was second only to Japan among Asian nations in terms of industrial output but since the 1970s its economic performance has lagged behind the "Asian Tigers" of Taiwan, South Korea, Hong Kong and Singapore who engaged in rigorous growth facilitated by export promotion (Balisacan 2003). The comparison with Taiwan is noteworthy; in 1962, the Philippines and Taiwan were roughly similar on many socioeconomic indices but by 1986, per capita income in the archipelago was only one-seventh of its neighbor to the north (Eder 1999). From 1990 to 2005, the growth of gross domestic product (GDP) per capita in the archipelago averaged only 1.6 percent a year; a number lower than Vietnam (5.9 percent), Laos (3.8 percent), Cambodia (5.5 percent), and even Myanmar (6.6 percent) (Bello et al. 2009). With such poor economic performance the Philippines have also become heavily indebted to external creditors; in 2008 the total value of all external debt was equal to 77 percent of all export revenue and debt service payments took up 16 percent of all export revenue (*Economist* 2011).

A landscape of poverty and marginalization

The most serious consequence of such poor economic performance has been the high levels of poverty pervading Philippine society. In 2006, the official poverty rate in the archipelago stood at 33 percent (World Bank

2010) while a study of self-rated poverty conducted that same year by the nongovernmental organization (NGO) Pulse Asia found that 74 percent of all Filipinos described themselves as poor (Tabunda 2007). While the distinction between the official poverty statistics and the self-rated statistics of poverty may be a noteworthy and interesting source of academic discussion there can clearly be no doubt that poverty is a serious problem in the archipelago. As Balisacan (2003, 311) wrote,

> A peculiar aspect of Philippine development in recent decades is the rather slow pace of poverty reduction and the persistently high level of economic inequality. Among the major East Asian economies, the Philippines has had the slowest rate of poverty reduction during the last three decades and, at the turn of the present century, had the highest incidence of absolute poverty.

Where poverty becomes particularly noticeable is in rural areas where people are primarily involved in agricultural activities. While, in 2006, the national poverty rate may have stood at 33 percent the rural poverty rate stood at 46 percent and 71 percent of all poor people lived in rural areas (World Bank 2010). In Figure 1.3 the 2005 rural poverty rates are presented on a province-by-province basis and rural poverty rates range from a low of 17 percent in Bataan up to a high of 72 percent in Sulu (National Statistical Coordination Board 2005).[5]

Spaces of vulnerability

The islands of the Philippines are also spaces of vulnerability as well as spaces of poverty. The principal source of this vulnerability is the heavy reliance of the rural population upon subsistence agriculture and subsistence aquaculture with 44 percent of those engaging in subsistence agriculture living in poverty and

---

5 The reader may note the high HDI score of 0.787 for the province of Benguet and the, seemingly incongruous, rural poverty rate of 48 percent for the same province. This is attributable to the fact that economic conditions for the province of Benguet as a whole are distorted by the presence of Baguio City within that province. Baguio City, "the summer capital of the Philippines," is a hot weather vacation center for the Philippine elite, home to a number of universities and colleges and it is also the location of the Philippine Military Academy. In 2005 Benguet had an overall provincial poverty rate of 25 percent but when only rural areas were looked at the poverty rate jumped 23 percentage points to 48 percent (National Statistical Coordination Board 2005). These rural poverty rates are calculated using a cost of basic needs approach wherein poverty lines are calculated to represent the monetary resources required to meet the basic needs of household members (National Statistical Coordination Board 2005).

Figure 1.3. Rural poverty rates by province

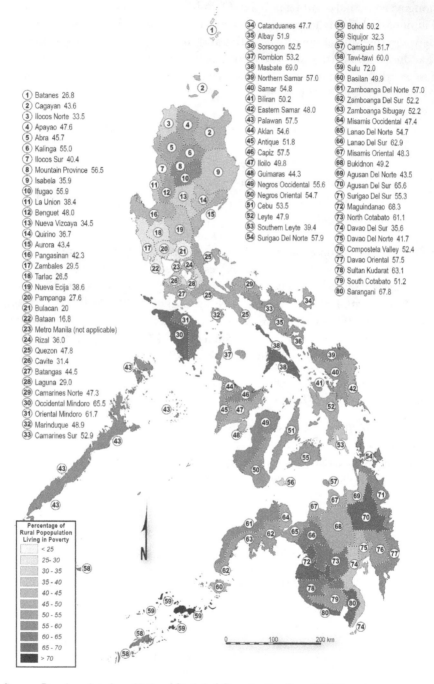

Source: Based on data from National Statistical Coordination Board (2005).

50 of those engaging in subsistence aquaculture living in poverty (Department of Environment and Natural Resources Climate Change Office 2010). People who engage in such activities are acutely vulnerable to any form of environmental disruption. As subsistence farmers and fisher-folk, poor people's livelihoods depend directly on ecosystems and the quality of and access to natural resources is paramount to their wellbeing. Across the archipelago, agriculture is the most important economic activity for the majority of people and the principal basis of economic activity (Bankoff 2003a). In the Philippines, agriculture accounts for about two-thirds of the labor force and 40 percent of the gross domestic product (Bello et al. 2009; Borras 2007; David 2003). Approximately 70 percent of the poor are based in the rural sector and are dependent directly upon agriculture-related economic activities for their major source of livelihood (David 2003). These people, engaging in subsistence activities, depend on natural resources for their existence; what could be referred to as "the ecological dependence of peasant livelihood" (Nadeau 1992, 58). Any degradation of the environment upon which subsistence farmers and fisher-folk depend will lead to adverse economic consequences that will be pervasive and profound (Myers 1988). As environmental quality decreases, these people will become poorer and their options for a better life will diminish (Roque and Garcia 1993). "Environmental deterioration is crucial to a resource-dependent society because it diminishes their ability to provide for their basic needs" (Roque and Garcia 1993, 89). Perhaps the best description of the importance of natural resources to the poor is that given by Broad (1994, 814):

> To live, poor people eat and sell the fish they catch or the crops they grow –
> and typically those who manage to subsist in this way do so with very
> little margin. Natural resource degradation often becomes an immediate
> and life- and livelihood-threatening crisis – a question of survival.

This vulnerability of the rural poor to environmental degradation exemplifies the relationship between the global and the local. Activities may be engaged in by powerful elements of society, such as multinational corporations, in response to global economic forces. However, should these activities degrade the environment relied upon by the rural poor the consequences for local communities reliant upon those resources will be catastrophic and those communities may never fully recover.

A landscape of violence

With so many people, particularly in rural areas, living in poverty the archipelago has experienced a substantial amount of unrest and this unrest has resulted in the presence of armed insurgent groups who defy the authority of the Philippine state. "Insurgency" is defined by Roque and Garcia (1993, 94) as "those movements that deny the authority of the existing Philippine government and are ready to use force to assert their demands." In the Philippines there are a number of different insurgent groups but two of them are most noteworthy: the Moro Islamic Liberation Front (MILF)[6] and the New People's Army (NPA).[7]

The Moro Islamic Liberation Front

The MILF (Figure 1.4), with an armed strength of 11,769 members (as of 2007), are the largest insurgent group in the Philippines (Santos and Santos 2010a). The MILF broke away from the Moro National Liberation Front (MNLF) in 1977 and, after the MNLF signed a peace agreement with the Philippine government in 1996, has been engaged in an on-again off-again struggle with the government. The MILF is an Islamist group that seeks a homeland for the Muslim (or "Bangsamoro") inhabitants of Mindanao and the Sulu islands. Whereas the MNLF was satisfied in having a degree of autonomy granted to the Muslim-dominated provinces[8] of the Autonomous Region of Muslim Mindanao (ARMM), the MILF is attempting to obtain an independent Islamic state in this part of the Philippines. As Santos and Santos (2010a, 344) wrote, "the MILF is a Moro rebel group with ambitions that are subnational geographically but national in relation to the Moro nation." The MILF operates almost exclusively in the ARMM and, on occasion, in portions of North Cotabato and Sultan Kudarat adjacent to the ARMM.

---

6 The Muslim inhabitants of Mindanao and the Sulu islands are called "Moros" because when the Spanish originally encountered them the Spanish were reminded of the Moors of North Africa. The Spanish originally referred to them as "Maurus" or "Mauris" and eventually came to call them "Moros" (Holden 2009b).

7 Another insurgent group which attracts substantial media attention, arguably out of proportion to its importance, is the Abu Sayyaf ("Bearer of the Sword") Group, or ASG. The ASG is a nominally Islamist group operating mainly in the Zamboanga Peninsula of Mindanao and in the Sulu islands. It is believed that the ASG was created by the AFP to generate internal conflict among the MNLF during the early 1990s (McCoy 2009). However, after the AFP created the ASG it became unmanageable and it morphed into a kidnap gang. The ASG consists of only around 650 fighters but its importance is overshadowed by the notoriety it has received from such incidents as its brazen May 2001 speedboat raid on the Dos Palmas resort off of the island of Palawan where twenty tourists were kidnapped.

8 Basilan, Lano del Sur, Maguindanao, Sulu, and Tawi-Tawi.

Figure 1.4. Moro Islamic Liberation Front members pray inside their mosque

Photo credit: Keith Bacongco.

The New People's Army

The most widespread insurgent group in the Philippines is the New People's Army (NPA), the armed wing of the Communist Party of the Philippines (CPP) (Figure 1.5). The CPP was reestablished along Maoist lines during the late 1960s by Jose Maria Sison (an English professor from the University of the Philippines) replacing the old Marxist-Leninist Partido Komunista ng Pilipinas (Communist Party of the Philippines, or PKP) on 26 December 1968, the 75th birthday of Mao Zedong (Sison 2007). The NPA was established by the CPP on 29 March 1969, the 27th anniversary of the founding of the Hukbalahap, and since then has been engaged in a conflict against the Armed Forces of the Philippines (AFP) in an attempt to overthrow the Philippine government and replace it with a communist regime (Santos 2010). This conflict has taken over 40,000 lives and is one of the longest running Maoist insurgencies in the world (Rutten 2008). Although estimates of the NPA's strength vary from between 5,000 (Hasting and Mortela 2008) up to 7,000 (Rutten 2008) and although many experts on the NPA regard it as being incapable of seizing power any time soon (Caouette 2004), there is widespread consensus that the NPA is the largest threat to the security of the state due to the fact it operates in all areas of the Philippines with the notable – and important – exception of

Figure 1.5. NPA cadres celebrate the 40th anniversary of the founding of the CPP

Photo credit: Keith Bacongco.

the ARMM (Caouette 2004; Espuelas 2008; Hasting and Mortela 2008; Rutten 2008).[9]

Figure 1.6 depicts the spatial distribution of encounters between the AFP and the NPA from 2000 to 2004[10] and one can see the widespread presence of NPA activity across the archipelago, particularly in the Southern Tagalog Region (Quezon Province), the Bicol Region (Albay), the Eastern Visayas (the island of Samar) and in Eastern Mindanao (Compostela Valley and Davao Oriental). Although the MILF has a numerical strength superior to that of the NPA, the Philippine government considers the MILF to pose less of a threat than the NPA because the MILF is geographically confined to one region of the country whereas the NPA operates nationwide (Santos and Santos 2010a).

Poverty: The common denominators of both insurgencies

Both insurgencies (Communist and Muslim) have the same common denominator: poverty. "The rebellions were, and still are, social and political protests against inequality, deprivation, oppression and injustice spawned by government bias for the rich and privileged in complete disregard for the poor and the underprivileged" (Diaz 2003, 150). "Given the widespread poverty," wrote Santos and Santos (2010b, 262), "there is no shortage of disenfranchised poor who might be recruited to the CPP and NPA." The NPA attracts two types of people: members of the rural poor (for whom it represents a genuine livelihood opportunity) and intellectuals and idealist who continue to find in its ideology a "clear, coherent, and realizable alternative to oligarchic politics and the abuses they see around them" (Santos 2010, 30).

Overseas Filipino workers: Modern heroes

With poor economic performance, a population experiencing high rates of poverty and violent insurgency occurring in the countryside, arguably the only thing keeping the Philippine economy operating are those Filipinos known as overseas Filipino workers (OFWs) who work abroad and then remit funds

---

9 The fact that the NPA does not operate in the ARMM is due to the fact that the NPA and MILF reached an agreement in 1999 not to operate in each others areas; the NPA does not operate in the ARMM and the MILF does not operate in Christian areas of Mindanao (Santos and Santos 2010a, 2010b).

10 Data on confrontations between the AFP and the NPA is provided by the IBON databank; IBON stopped decomposing its data on a province by province basis at the end of 2004.

Figure 1.6. Confrontations between the AFP and the NPA, 2000–2004

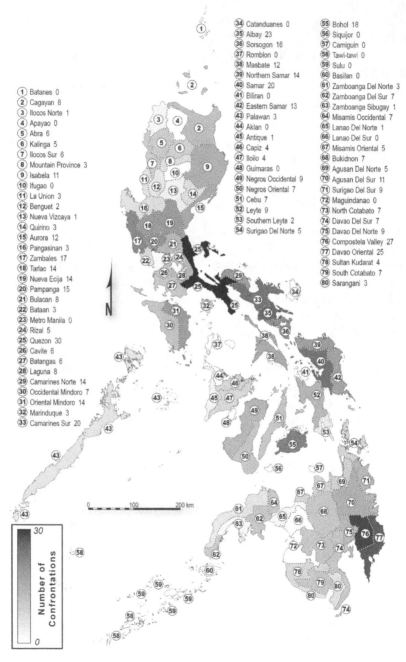

Source: Based on data from IBON (2001a, 2001b, 2002a, 2002b, 2003a, 2003b, 2004a, 2004b, 2005).

home to their families (Rodriguez 2010; Tyner 2009). From across the archipelago, at any point in time almost ten percent of the population work abroad temporarily at a variety of occupations such as bartenders, construction workers, maids, nannies, performers, seamen, waiters and waitresses (Rodriguez 2010; Tyner 2009). In 2008, the Philippines was fourth in the world (behind India, China and Mexico) in remittances from citizens working abroad and almost USD 19 billion were remitted from OFWs back to their families in the Philippines (*Economist* 2011). Overseas Filipino workers have become ubiquitous around the world and the government has eagerly encouraged Filipinos to work abroad.[11] As Tyner (2009, xiv) wrote, "The Philippines has become the world's largest exporter of government sponsored temporary contract labor." The remittances of OFWs are a major source of revenue that the state is eager to receive and OFWs are touted as "modern heroes" who perform a patriotic service (Rodriguez 2010; Tyner 2009). Indeed, the government openly admits the vital role overseas employment serves in acting as a safety valve in preventing social unrest within the country (Rodriguez 2010). If otherwise unemployed Filipinos are working abroad there will fewer recruits for the NPA and OFW remittances will mitigate the misery generated by poverty and make the NPA's protracted people's war seem less appealing to their family members in the Philippines (Rodriguez 2010).

The heavy reliance of the Philippines also demonstrates the relationship between the local and the global. Should Filipinos find themselves unable to find livelihoods at home due to economic conditions in the Philippines, they will become part of what Rodriguez (2010, 53) calls the "reserve army of Philippine labor at the ready for deployment around the world." Poor local economic conditions in the archipelago will increase the number of Filipinos ready and willing to become OFWs and this will supply global labor markets with more people to work as bartenders, construction workers, maids, nannies, performers, seamen, waiters and waitresses. Indeed, the global and the local can feed back on each other. Activities done in the islands in response to global market forces may degrade the environment relied upon by the rural poor. The rural poor, denied access to resources necessary for subsistence activities where they reside, will then be forced into the reserve army of Philippine labor

---

11 An excellent example of how the government has aggressively promoted overseas employment is its hosting of delegations from the four western Canadian provinces of Alberta, British Columbia, Manitoba and Saskatchewan (Rodriguez 2010). In Canada, OFWs have established a ubiquitous presence working in Tim Horton's, the iconic Canadian chain of coffee shops named after Tim Horton (a former ice hockey player). Tim Horton's is considered an archetypal aspect of Canadian culture, akin to toques, maple syrup and ice hockey. Nevertheless, most Tim Horton's locations in the four western provinces are staffed by OFWs working in Canada on temporary work permits.

and made available for the global labor market; the global effects the local which then feeds back upon the global.

Mineral Resources of the Philippines

Given the high (and persistent) rates of poverty pervading the archipelago, the government has viewed the mineral resources of the Philippines as a method of accelerating its economic performance. Starting in the 1990s, this accelerator has increasingly been seen to lie below the surface of the islands. The same geologic forces that have caused the archipelago to become so mountainous have also endowed it with rich mineral resources (Lopez 1992). Geologists divide mineral resources into two categories: metallic mineral resources and nonmetallic mineral resources (Table 1.1). The metallic mineral resources are then divided further into abundant metals (such as iron), with a crustal abundance above 0.1 percent, and the scarce metals (such as gold), with a crustal abundance below 0.1 percent (Skinner 1976). Since iron is not included among the scarce metals, the scarce metals are consequently referred to as the nonferrous metals.

Since nonferrous metals are frequently (although not exclusively) found in consolidated rock of igneous or metamorphic origin they are referred to as hardrock minerals and the mining of these minerals – which will involve considerable crushing and pulverization – is referred to as hardrock mining (Francaviglia 2004). Given that many of the scarce minerals are often found together in combinations (such as copper and gold, or lead and silver, or copper, lead, silver and zinc) they are often referred to as polymetallic deposits. Copper, for example, is typically found in conjunction with gold, lead, silver and zinc (Peck et al. 1992). This means that when mining, companies engage in exploration; they are searching for a collection of minerals rather than just one mineral (Peck et al. 1992).

Table 1.1. Mineral resources

Metallic mineral resources	Nonmetallic mineral resources
Abundant metals: iron, aluminum, manganese, titanium, and magnesium	Minerals for chemical fertilizer, and special uses: sodium chloride, phosphate, nitrates, and sulfur
Scarce metals: copper, lead, zinc, tin, tungsten, chromium, gold, silver, platinum, uranium, mercury, and molybdenum	Building materials: cement, sand, gravel, gypsum, asbestos
	Fossil Fuels: coal, petroleum, natural gas, and oil shale
	Water: lakes, rivers, and ground waters

Source: Skinner (1976).

The archipelago is located on the Philippine Plate, a rigid portion of the earth's outer crust referred to as the lithosphere (Punongbayan 1994). As the heat generated in the earth's core rises into the upper portion of the earth's mantle, tectonic plates (portions of the lithosphere), move over geologic time upon the upper mantle (Rimando 1994). When two of these plates converge with each other, one plate will be subducted and move below another plate (Rimando 1994). The Philippine Plate is located between the eastward moving Eurasian Plate, the northward moving Indo-Australian Plate and the westward moving Pacific Plate (Figure 1.7). As the Eurasian Plate, the Indo-Australian Plate and the Pacific Plate are subducted beneath the Philippine Plate, the rocks of these plates are heated and turned into magma which – having a lower density than the surrounding rocks – rises towards the surface (Punongbayan 1994). As this magma rises, it creates island arcs and the islands of the Philippines are an example of such an island arc (Mitchell and Leach 1991). The rising magma also creates volcanic mountains and endows these mountains with mineral deposits (Lopez 1992).

Figure 1.8 depicts a block diagram of an idealized granite structure. During the mountain building (or orogenic) process a body of molten igneous rock moves upward through the earth's crust; around this body of molten rock is heated water carrying dissolved metals. As this body of molten igneous rock rises through the surrounding rock it will create cracks, or fissures, in this surrounding

Figure 1.7. The world's tectonic plates and the Pacific Ring of Fire

Figure 1.8. Block diagram of an idealized granite structure

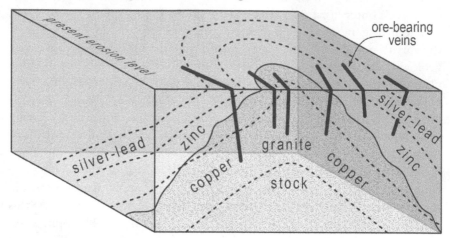

rock and water will flow through these fissures. As this water approaches the surface of the earth, it cools and precipitates metals into the fissures through which it traveled and these fissures will constitute ore-bearing veins. This process of creating and depositing minerals is referred to as hydrothermal, or epithermal, mineralization (Lopez 1992; Mitchell and Leach 1991).

With a history of volcanic activity, the Philippines is well endowed with ore deposits created through epithermal mineralization (Lopez 1992; Mitchell and Leach 1991). Although the mineral deposits of the archipelago are not exceptionally high-grade deposits, there tend to be numerous mineral deposits located throughout the archipelago (Lopez 1992; United States Bureau of Mines 1991). In the 2010/2011 Fraser Institute[12] Annual Survey of Mining Companies, 68 percent of the 494 respondents view the mineral potential of the Philippines as a strong factor encouraging investment in the country (Fraser Institute 2011).

Most of the mineral deposits in the Philippines are of nonferrous metals and these have historically accounted for 75 percent of the production value and nearly 100 percent of all mining export revenue (United States Geological Survey

---

12 The Fraser Institute is a neoliberal "think tank" located in Vancouver, British Columbia, Canada, which advocates a minimal amount of government interference in society as a generic solution to social problems. Starting in 1998, the Fraser Institute began surveying mining companies on their opinions of various jurisdictions as possible locations of mining investment. While the methodologies and accuracy of the Fraser Institute report are not without controversy, it is one way of taking a broad look at the mining industry worldwide.

1996). The value of the mineral resources of the archipelago is estimated at USD 1.4 trillion (Joint Foreign Chambers of the Philippines 2010). Nationwide, it is estimated that there are 4.8 billion tons of copper deposits and 111,000 tons of gold deposits (Carreon 2009). There are abundant deposits of gold in the Cordillera of Luzon, on Leyte and on Mindanao. Copper is found on Luzon, on Cebu and on Mindanao; chromium is found on Luzon, Samar and Mindanao and nickel can be found on Mindoro, Palawan and Mindanao (Lopez 1992).

Something that greatly eases any mining operation carried out in the archipelago is the fact that virtually every point on even the largest of the islands is no more than 100 kilometers away from the coast (United States Bureau of Mines 1991). Consequently, mines will always be in close spatial proximity to deep-water anchorages and ships can easily be loaded for exporting minerals from the islands. This is an important observation since countries with coastal access are consistently preferred to landlocked countries in the mining industry decisional calculus (Naito et al. 2001).

The History of Large-Scale Mining in the Philippines

The Philippines has a long mining history dating back almost a thousand years (Lopez 1992). Gold mining dates back to the third century AD[13] when Chinese merchants referred to Luzon as "the Island of Gold" (Rovillos et al. 2003) and by the fourteenth century AD, crudely smelted copper was traded to the Chinese (Mines and Geosciences Bureau 2000). Gold mining began in 1570 in the Bicol Region of Luzon during the Spanish colonial period (1565–1898) (Mitchell and Leach 1991). However, it was not until the American colonial period (1898–1946) that large-scale mining began in earnest.

In June of 1898, representatives of the United States Geological Survey were sent to the islands along with the soldiers of the United States Army "to secure information regarding geological and mineral resources" (Constantino 1975, 301). According to (Lopez 1992, 46), "The geologists who joined the American expeditionary forces in the late 1890s produced a mineral map of the archipelago that supported their assessments that bright prospects existed for a viable mining industry in the Philippines." Early reports back to the United States from these geologists embedded among American military forces "read like a mining stock prospectus" (Constantino 1975, 301). As the Americans began transforming the Philippines from a spoil of the Spanish–American War into a foreign colony, they also began to engage in a systematic

---

13 The authors admit the Judeo-Christian centric use of the terms "BC" and "AD" but employ these out of familiarity with their use and because of the prevalence of these terms in literature written in the Philippines, an overwhelmingly Christian country.

extraction of its mineral resources and in 1900, even before the Philippine–American War had concluded, Lieutenant Charles H. Burrett (an army officer) was appointed Chief of the Bureau of Mines established by the Insular Government (Lopez 1992). By the end of the first decade of the twentieth century, American prospectors (mostly former soldiers) had staked claims in the Baguio district of Benguet Province in the Cordillera (Lopez 1992). In 1903 the Benguet Corporation was established and in 1905 it commenced commercial mining in the Baguio area (Mitchell and Leach 1991). Between 1907 to 1914 Warren D. Smith, chief of the Division of Mines, Bureau of Science, recorded the discovery of chromium on Luzon, Mindoro, Panay, and Mindanao (Mines and Geosciences Bureau 1999). By the 1930s, mining in the archipelago had become a well-established activity and "Baguio was transformed from a vast wilderness into a bustling metropolis" (Tujan and Guzman 2002, 34).

During the American colonial period, the growth of large-scale mining went through three distinct phases: 1899–1919, 1920–1930 and 1930–1942 (Lopez 1992). The first phase was associated with the rapid technological change occurring during the late nineteenth century that facilitated large-scale industrial mining; the second phase was characterized by the first local gold boom precipitated by the sensational gold finds in the Cordillera that attracted many veteran American prospectors to the islands; the third phase was the era of the second gold boom brought about, paradoxically, by the Great Depression. During the early years of the Great Depression (1930–33) the price of gold was fixed by the United States government as part of its adherence to the gold standard and this meant that the price of gold was stable at a time when all other commodity prices were falling rapidly (Lopez 1992). Then, when Great Britain went off the gold standard in 1931 and the United States followed suit in 1933, the real price of gold rose by 76 percent between 1931 and 1934 (United States Geological Survey 2011). By 1935 the Philippines produced more gold than any state in the United States with the exception of California and the total value of gold exports from the islands was exceeded by only sugar and coconut (Lopez 1992). By 1941, the Philippines ranked as the world's fifth largest gold producer (Oliveros 2002). The gold booms of the 1920s and 1930s occasioned extensive mineral exploration throughout the archipelago that led to substantial discoveries of base metals (such as copper and nickel) and by the late 1930s, as war preparations bid up base metals prices, these minerals began to experience a boom of their own (Lopez 1992).

The Japanese invasion and occupation (1942–45) disrupted mining substantially and damage to mining infrastructure accounted for 21 percent of all damage to major industries (Lopez 1992). However, the post–World War II reconstruction period led to a substantial demand for copper and the 1950s and 1960s saw the boom of the copper mining industry (Lopez 1992). By 1970, Atlas Consolidated Mining's Toledo Mine (on the island of Cebu) had become the largest copper mine in Asia and by 1971, it had become one of the ten largest copper mines in the world (United States Bureau of Mines 1970, 1971). By 1974, the Philippines was home to 18 major copper mines and by 1980 copper production peaked at over 304 thousand metric tons of copper while 45 large-scale mines were responsible for over 20 percent of all exports from the archipelago (Rovillos et al. 2003). However, during the turbulent decade of the 1980s (with its political unrest, worldwide global recession and falling mineral prices) a progressive decline of the mining industry in the islands began and 30 large mining operations shut down during that decade (*Mining Journal* 1995). The percentage of export earnings attributable to mining fell from over 20 percent in the early 1970s to only 6 percent in 1992 (*Mining Journal* 1995). By the early 1990s, the mining industry began to become viewed as a woefully underutilized agent of economic development (Rovillos et al. 2003). "The industry," wrote Lopez (1992, 324), "has become a wretched picture of a comatose enterprise, if it is not already in a state of rigor mortis." In 1993 the London based *Mining Journal* published an article entitled "Philippines' Mining Headed for Collapse?" and this article stated, "The Philippines' mining industry is lagging far behind that of other Southeast Asian economies" (*Mining Journal* 1993, 89). It was during this time that the Philippine government began to engage in substantial – and concerted – efforts to encourage foreign direct investment into mining. However, to discuss these efforts properly, attention must turn to the intellectual impetus behind the promotion of mining investment: neoliberalism.

Neoliberalism is a theory proposing the advancement of human welfare through the liberation of entrepreneurial freedoms within "an institutional framework characterized by strong property rights, free markets and free trade" (Harvey 2005, 3). Neoliberalism may be equated with "a radically free market, maximized competition and free trade achieved through economic deregulation, elimination of tariffs and a range of monetary and social policies favorable to business" (Brown 2003, 1). As the neoliberal paradigm began to spread throughout the world in the late 1970s and early 1980s the role of the state began to change from being one of a direct facilitator of economic activity – through aggressive interventionist policies (and even outright government ownership of enterprises) – to that of being an indirect facilitator of economic activity, which created a stable investment regime conducive to the entry of multinational corporations.

To the traditional holders of power in many societies neoliberalism appears highly attractive as it rejects income redistribution as inefficient and views the poor as those who have failed to give their lives proper entrepreneurial shape (Brown 2003). Consequently, neoliberalism is often described as "an international project to reclaim, reconstitute, or establish capitalist class privilege and power" (Heynen et al. 2007, 290). The focal point for neoliberal theorists is not workers in factories or peasants on plantations. To them the "free individual" means the entrepreneur, the capitalist and the boss; to them "freedom" means the opportunity to make money (Peet and Hartwick 2009). Given the scope inherent in neoliberalism for an expansion of upper-class power, it has become quite successful in the Philippines, a nation dominated by a powerful oligarchy where "the state has been an instrument for class domination" (Hawes 1987, 131). "The Philippine state," wrote Hawes (1987, 133), "operates in the interests of the bourgeoisie." Neoliberalism, a paradigm expressly eschewing wealth and income redistribution while also calling for a strong maintenance of law and order so as to preserve property rights, has found fertile ground in the Philippines as the archipelago has long been dominated by a wealthy elite that has controlled the coercive apparatus of the state to protect its interests. The genesis of neoliberalism in the universities of the developed world – and its subsequent diffusion to the developing world – must now be examined.

Chapter Two

GOVERNMENT EFFORTS TO ENCOURAGE MINING

The Ascendency of Neoliberalism

Modernity: Trust and confidence in experts

The word "modern," in its Latin form *modernus*, was first used in the late fifth century in order to distinguish the (then) present, which had become officially Christian, from the Pagan past (Habermas 1981). Modernism is infused with a belief (inspired by science) in the infinite progress of knowledge and in the infinite advancement of humans towards social betterment. "Generally perceived as positivistic, technocentric and rationalistic, universal modernism," wrote Harvey (1990, 9), "has been identified with the belief in linear progress, absolute truths, the rational planning of ideal social orders and the standardization of knowledge." In large part, modernism is centered on "the rationality of human beings, the privileged status of science as the only valid form of knowledge, the technological mastery of nature [and] the inevitability of progress" (Warf 1993, 162). An essential component of modernity is trust and confidence in experts. As Escobar (1996, 55) wrote: "One of the defining features of modernity is the increasing appropriation of 'traditional' or pre-modern cultural contents by scientific knowledge and the subsequent subjection of vast areas of life to regulation by administrative apparatuses based on expert knowledge."

Economics as a discourse of modernity

The trust and confidence in experts, which is a hallmark of modernity, applies very much to capitalism. The history of capitalism is intimately linked with the advancement of scientific discourses of modernity in the discipline of economics. According to Escobar (1996, 55), "the modern form of capital is inevitably mediated by the expert discourses of modernity." Perhaps the best example of how the discipline of economics has come to be dominated by trust and confidence in experts is the statement from an unnamed top official of a leading international institution, quoted in Broad and Cavanagh

(1993, 154), "The world knows much better now what economic development policies work and what policies do not. Now we almost never hear calls for alternative strategies based on harebrained schemes." Today in economics, the dominant intellectual paradigm unquestionably established as the scientifically determined truth is neoliberalism. To Peet (2003, 4), "Neoliberalism has achieved the status of being taken for granted or, more than that, has achieved the supreme power of being widely taken as scientific and resulting in an optimal world." Indeed, neoliberalism has become so hegemonic that protest against it has come to be seen as an "offence against reason, progress, order, and the best world ever known to man" (Peet 2003, 4).

The origins of neoliberalism

The Great Depression of the 1930s was a profoundly influential period with respect to economic policy in the industrialized world. Many viewed the Great Depression as evidence that capitalism, left on its own, was incapable of providing economic progress and that a large degree of government involvement in the economy was necessary in order to ensure prosperity. One of the foremost proponents of such thinking was the British economist John Maynard Keynes who, in his 1936 book *The General Theory of Employment, Interest and Money*, established a theoretical basis for policies using government spending so as to maintain adequate demand in the economy for all of the products produced by it. Keynesian economics became highly influential and it dominated economic discourse throughout the 1940s, 1950s, 1960s and well into the 1970s. Economic textbooks written during this time period demonstrate how thorough the grip of Keynesian thinking was upon the discipline of economics. Consider, for example, the classic textbook *Economics: an Introductory Analysis*, written by the famous economist Paul Samuelson. This book which first appeared in 1948 was resplendently full of graphical illustrations demonstrating how increased government spending will close a "recessionary gap" and maintain "full employment" and it provided a generation of undergraduate students with a solid grounding in the importance of a highly interventionist government. Indeed, Keynesian economics became so successfully established that by 1971 President Richard Nixon (a Republican) stated "we are all Keynesians now" (Harvey 2005, 13).

However, during the "stagflation" (high inflation concomitant with stagnant growth) of the mid to late 1970s these policies began to appear weak and ineffective. Keynesian economic theory developed during the high unemployment of the 1930s could do a reasonably good job of explaining how to address high unemployment (by increasing government spending) but it encountered substantial difficulty explaining how to address high unemployment occurring simultaneously with high inflation. During the

1970s, socialist movements such as the communist parties of Italy and Spain began to appear asking for even more state involvement in the economy – such as wage and price controls and even bank nationalizations (Harvey 2005, 2006). To preempt these movements and to placate those with vested interests threatened by even more state involvement in the economy, neoliberalism emerged (Harvey 2005).

The term "neoliberalism" refers to a set of economic polices emphasizing free trade, privatization, deregulation and the retreat of the state from matters of wealth redistribution and social service provision (Peet 2003; Ward and England 2007). As Tyner (2009, 147 wrote):

The central features of neoliberalism include the primacy of economic growth, the importance of free trade to stimulate growth, the promotion of an unrestricted free market, individual choice, privatization of business, the reduction of government regulation, and the advocacy of an evolutionary model of social development anchored in the Western experience and applicable to the entire world.

The prefixing of the word "neo" onto the word "liberalism" is done to indicate that neoliberalism is a revival of the teachings of the classical liberals (such as Adam Smith) who articulated a view that social order will emerge as the consequence of many people each seeking their own interests (Harvey 2005; Ward and England 2007). As Vanden and Prevost (2006, 164) wrote, "Classical liberals believe that the magic hand of the market, not government control or trade barriers should regulate the economy." To Peet (2003, 8), neoliberalism "is an entire structure of beliefs founded on right-wing, but not conservative, ideas about individual freedom, political democracy, self-regulating markets and entrepreneurship." According to Ballve (2006, 27), neoliberalism "is used as the shorthand for the process of global capitalist restructuring that began in the 1970s in which the extremist ideology of the 'free market' was given pride of place as the basis for the whole organization of society."

There were a number of theorists behind neoliberalism but two economists from the University of Chicago, Friedrich von Hayek and Milton Friedman, stand out as the two most influential with their seminal books: *The Road to Serfdom* (Hayek 1944), *The Constitution of Liberty* (Hayek 1960), *Capitalism and Freedom* (Friedman 1962) and *Free to Choose* (Friedman and Friedman 1979).[1] Through these books their authors laid out a highly persuasive case for reducing the role of the state in the economy. While Hayek and Friedman may have been

---

1 Another highly influential progenitor of neoliberalism was the Austrian economist Ludwig von Mises who, in his Mont Pelerin Society, advocated freedom of expression, free market economic policies and the political values of an open society.

the two great theorists behind neoliberalism, its practical implementers were
Margret Thatcher (the prime minister of the United Kingdom from 1979–90)
and Ronald Reagan (the president of the United States from 1980–89). During
their time in power there was a substantial reduction in the involvement of the
state in the economies of both countries. In the United Kingdom, this was marked
by the privatization of state owned enterprises such as British Steel, British Rail
and British Telecom and a weakening of the power of the labor movement as
demonstrated by the violent mineworkers' strike of the 1980s. Thatcher's role
in transforming the United Kingdom was so thorough that by 1997, when
the Labour Party returned to power, it barely altered any of her changes and
Prime Minister Tony Blair "could easily have reversed Nixon's earlier
statement and simply said 'We are all neoliberals now'" (Harvey 2005, 13).

Neoliberalism in the Developing World

Structuralism: The precursor to neoliberalism

Just as the Great Depression (by encouraging Keynesian economics) influenced
economic policy in the developed world, it also influenced economic policy in
the developing world. During the 1930s as the Depression reduced international
trade (and during the 1940s as World War II disrupted international trade),
many developing nations were forced to try and develop their economies
on their own without access to markets in developed countries. In the late
1940s, after having observed the behavior of his own country's economy
during the 1930s and early 1940s, the Argentine economist Raul Prebisch
developed a model of economic development known as "structuralism" or
"import substitution." According to structuralism, the necessary condition for
development is to enhance the industrial structure of an economy; this is to be
done by erecting tariff barriers forcing foreign manufacturers to build branch
plants within the territory of a developing nation and by direct involvement
in the economy through government owned corporations. As these branch
plants and government owned enterprises are built, imports will be replaced by
domestically produced goods and the structure of the economy will become more
industrialized and, ultimately, developed. During the 1940s and 1950s, many
developing counties were heavily influenced by Prebisch and enthusiastically
pursued import-substitution industrialization policies (Broad 1988).

The role of the World Bank

Starting in the 1960s, however, some developing countries began moving
away from import substitution and started adopting policies emphasizing

export promotion; the crucial actor in this transformation was the World Bank (Broad 1988). Conceived at the 1944 Bretton Woods conference, which saw the establishment of the postwar international monetary regime, the World Bank was an institution created to provide financial and technical assistance to developing countries. During the 1960s, the World Bank began to have influence upon developing country governments who had recently seen military coups (often aided by the United States) replace nationalist leaning governments engaging in import substitution with transnational governments engaging in export promotion. This was the case in Brazil, in 1964, in Indonesia, in 1965 and in Chile after 1973 (Broad 1988). The involvement of the World Bank with the government of General Augusto Pinochet in Chile was quite noteworthy. From 1970–73, while Salvador Allende, a Marxist, was the democratically elected president of Chile, the World Bank constantly denied any aid to Chile (Spooner 1994). However, after the Chilean military (acting with substantial assistance from the United States) overthrew Allende on 11 September 1973, it initiated a series of radical free market reforms and the World Bank resumed lending to Chile (Spooner 1994).[2] By the late 1970s and early 1980s, the World Bank began imposing neoliberal polices upon developing countries by attaching conditions to the loans it made to them. These conditions typically involved changes to tariffs, tax rates, subsidies interest rates and public spending by the recipient nations. Eventually, after having received a steady diet of free market advice from the World Bank, and after having seen the ascendency of neoliberalism in the developed world, almost all of the countries in the developing world restructured their economies in accordance with the precepts of neoliberalism (Bello et al. 2009). Since the World Bank is headquartered in Washington DC, this acceptance of neoliberalism became known as the Washington Consensus (Vanden and Prevost 2006). Then, as the Berlin Wall collapsed in 1989, many writers began declaring "history to be over" and used the failure of Soviet socialism as evidence to demonstrate that all forms of state intervention in an economy are bound to fail (Fukuyama 1989). "Champions of neoliberalism," wrote Ellner (2006, 95), "pointed to the fate of Soviet socialism as hard proof that all forms of state intervention in the economy were doomed to failure." Advocates of neoliberalism "have argued that in today's global economy, the assertion of national sovereignty by strong third world governments has no potential for transformation" (Ellner 2006, 100). Economic policy, in most of today's developing countries, is focused on developing policies so as to create

---

2 This illustrates how the World Bank has long been criticized for failing to adequately take into account the human rights policies of the governments to which it lends money (Broad 1988).

conditions permitting multinational corporations to enter markets (De Rivero 2001). The role of the state, under neoliberalism, is that of fostering a positive investment climate for multinational corporations; these entities roam the world looking for the lowest cost locations in which to locate their operations by selling their products in global markets, acting under the control of no one government and accounting only to their shareholders.

Because of the importance placed by neoliberalism on attracting multinational corporations into developing countries, a term frequently used in conjunction with neoliberalism is globalization. Globalization is the tendency for economic interdependencies to occur on a global scale. Although activities have occurred on a global scale for years, neoliberalism, with its heavy emphasis on free trade, has led to such an amplification of globalization that Ward and England (2007, 12) have taken to calling "globalization" the "international face of neoliberalism." Indeed, globalization and neoliberalism have a powerful synergy between them; the globalization of the world economy is often used to justify the implementation of neoliberal economic policies wherein an interventionist state is viewed as a hindrance to economic development (Tyner 2009). Globalization provides authority to neoliberal arguments for unimpeded access to markets and freer trade and "creates the context in which openness to the global market is seen as the inevitable and commonsensical route to prosperity and progress" (Tyner 2009, 33).

Neoliberalism and Mining

Just as the World Bank has encouraged foreign direct investment, in general, as a development strategy, it has also specifically encouraged foreign direct investment into mining as a development strategy (Bebbington et al. 2008). This emphasis upon mining has been pursued to quicken economic growth and to earn foreign currency so as to comply with the conditions of loans made by the World Bank to these countries (Yocogan-Diano et al. 2009). Since 1985, more than ninety states have adopted new mining laws, or revised existing ones, in an effort to increase (and in some cases, initiate) foreign investment in the mining sector of their economies (Bridge 2004, 2007). According to Pegg (2006, 383):

> The World Bank's approach to mining and poverty reduction throughout most of the 1980s and 1990s could be characterized as a "Field of Dreams" approach. As in that movie, the World Bank Group's vision was guided by the idea that "if you build it, they will come." In this case, the Bank was trying to attract foreign investment from transnational mining firms. What they built to attract this foreign investment was an investor-friendly regulatory regime.

This thinking began to become apparent in two highly influential World Bank publications that emerged during the early 1990s (Moody 2007; Rovillos et al. 2003). In 1992, the World Bank published *The Strategy for African Mining* (World Bank 1992). This report argued against public ownership of mining companies in developing countries and placed an emphasis on mining projects that would be aimed primarily at production for export instead of providing raw materials for the domestic economy (Moody 2007). Two years later, in 1994, John Strongman, the principal mining officer of the World Bank, published *Strategies to Attract New Investment for African Mining* (Strongman 1994). This report made it clear that the World Bank's emphasis was on opening up the mining sector in developing countries to foreign direct investment from multinational mining companies and on improving the efficiency and competitiveness of mining projects (Rovillos et al. 2003). Although these two reports were designed to stimulate the economic development of Africa, a continent of extensive poverty underlain by substantial mineral wealth, they had a substantial impact on the crafting of national mining codes not only in Africa but in other parts of the developing world as well (Moody 2007; Rovillos et al. 2003). "Most governments all over the world," stated Yocogan-Diano et al. (2009, 2), "are aggressively promoting the mining industry and supporting the interest of transnational mining companies." This has been done within the framework of mining liberalization that the World Bank has advocated since the early 1990s. Often such policies are implemented to comply with conditions imposed by the World Bank on loans it makes to developing countries. When the World Bank provides development assistance to countries, it will often say that such assistance will only be made if certain "conditions" – such as changes to a legal system to facilitate more foreign direct investment – are implemented by the recipient nation; this is often referred to as "World Bank loan conditionality." As Yocogan-Diano et al. (2009, 2) wrote about Asia:

> Most Asian governments [have] enacting mining liberalization laws and [enabled] national mineral policies to benefit mining corporations by granting them wide incentives. This is also a quicker way for the governments to increase gross domestic product and foreign currency to meet the conditionality attached to the World Bank's lending.

Neoliberalism in the Philippines

Import substitution industrialization during the 1950s

By the late 1940s the Philippines began to suffer from a growing trade deficit caused by decreased American spending in the archipelago, declining prices for agricultural exports and increasing imports of consumer goods (Hawes 1987). These problems caused the government to engage in structuralism and in 1948 the Import Control Act (Republic Act 330) was passed (Tyner 2009). The passage of this statute, along with the imposition of exchange controls in 1950 and the erection of high tariff barriers, "resulted in a period of import substitution industrialization that lasted for more than a decade" (Hawes 1987, 33). At this time, President Carlos Garcia (1957–61) advanced a "Filipino First" policy and throughout the Garcia presidency import substitution was dominant, building much of its strength on an appeal to nationalism (Hawes 1987). However, by the early 1960s, the value of Philippine exports was falling and the economy began experiencing a slowdown in growth (Tyner 2009). With many manufacturing industries unable to expand beyond the limited protected domestic market, the limited economic growth proved incapable of absorbing the expanding labor force and structuralism began to fall out of favor (Tyner 2009).

Export promotion under Diosdado Macapagal

"The 1962 election of Diosdado Macapagal," wrote Broad (1988, 33), "brought to power an administration pledged to terminate a decade of import substitution industrialization protected by exchange and import controls." Macapagal was elected with strong support from agricultural exporters who expected him to lift import and exchange controls which they viewed as an impediment to their operations (Hawes 1987). Much of Macapagal's plan for stimulating economic growth rested on increasing agricultural exports and the way to do this was to attract foreign investors willing to engage in agricultural production for export. Macapagal was determined to develop a laissez-faire economy with a limited role for government. While this policy stimulated new agricultural investments and a growth in agricultural exports, it also resulted in a slowdown in industrialization.

Ferdinand Marcos and enhanced export promotion

One of the most vocal opponents of import substitution industrialization during the Garcia presidency was Senator Ferdinand Marcos (Putzel 1992). When Marcos was elected president in 1965, reelected in 1969 and ensured of virtually absolute powers by declaring martial law in 1972, the stage was set for a heavy promotion of export-oriented industrialization. Development policy during the Marcos years became increasingly oriented toward the production of exports and economic policies were primarily designed to attract new private investment in export producing industries (Tyner 2009). The Investment Incentives Act

(Republic Act 5186) – passed in 1967 – and the Export Incentives Act (Republic Act 6135) –passed in 1970 – further opened the economy to foreign investors and in 1972, Presidential Decree 66 was issued creating export processing zones where unassembled parts could be imported duty free for the purposes of assembling manufactured goods such as electronics for export (Hawes 1987). In 1969 the first export-processing zone was established in Bataan; later in the 1970s other export processing zones were established in Baguio City and on Mactan Island near Cebu City (Tujan 2007). Martial law (1972–81) was a boon to manufacturing exports since the power and wages of the working class were sharply curtailed through strict surveillance of unions and due to draconian restrictions on the right to strike (Hawes 1987). The Marcos presidency also saw the beginnings of the archipelago's heavy reliance on overseas Filipino workers as Marcos "saw the export of labor as an important measure to curb the political unrest likely to be exacerbated by unemployment and underemployment" (Rodriguez 2010, 12).

Fidel Ramos: Philippines 2000

It was, however, during the presidency of Fidel Ramos (1992–98) that the Philippines began to get its reputation of being among the most compliant in all of Asia to the prescriptions of neoliberalism (Quimpo, N.G. 2009). Ramos began implementing a rigorous program of neoliberal reforms entitled "Philippines 2000," which aimed to make the Philippines a "developed country" by the year 2000 (Nadeau 2002, 2008). Ramos and members of his administration saw the high growth rates of countries such as Hong Kong, Taiwan, Singapore and South Korea as products of free-market policies instead of strategic state interventions in the market (Bello et al. 2009). "Indeed, under Ramos a program of liberalization, deregulation and privatization was pursued with almost messianic zeal" (Bello et al. 2009, 92). To the Ramos administration, the ideal template for development was the Pinochet regime in Chile (1973–90) and "the administration's economists used Chile as a model, going so far as to invite the finance minister of the Pinochet dictatorship, Rolf Luders, to speak before many Filipino audiences" (Bello et al. 2009, 96). Ramos continued the aggressive promotion of exports established during the Marcos years and in 1994, the Exports Development Act (Republic Act 7844) was signed into law. This statute made it an express policy of the government to champion exports as a focal strategy for industrial development. Ramos further emphasized the reliance on OFWs initiated during the Marcos years (Rodriguez 2010; Tyner 2009). Clearly, by the end of the Ramos presidency in 1998, neoliberalism had become the dominant ideological paradigm in the Philippines (Bello et al. 2009).

Gloria Macapagal-Arroyo: The ideal neoliberal subject

One person who was instrumental in aiding Ramos implement his neoliberal agenda was (then) Senator Gloria Macapagal-Arroyo, the daughter of former president Diosdado Macapagal. A Georgetown University trained neoliberal economist, Macapagal-Arroyo's "experience in politics conformed to the routine career of a member of local oligarchic dynasties" (San Juan 2006, 7). While a member of the Senate, Macapagal-Arroyo sponsored several neoliberal reforms: the Senate ratification of the General Agreement on Tariffs and Trade and the Philippine accession to the World Trade Organization (Senate Resolution No. 97); laws facilitating comprehensive foreign investment liberalization (Republic Act 8179); banking law reform (Republic Act 7721); laws creating special economic zones (Republic Act 7916) and oil industry deregulation (Republic Act 8479) (International Coordinating Secretariat of the Permanent People's Tribunal and IBON Books 2007). When Macapagal-Arroyo (elected as vice president in 1998) replaced President Joseph "Erap" Estrada (after the latter's resignation amid allegations of widespread corruption) in January of 2001, neoliberalism achieved supremacy in the archipelago's policymaking process and Macapagal-Arroyo vigorously continued all of the neoliberal programs establish during the Ramos presidency. Macapagal-Arroyo viewed herself as not merely the president of the Republic but also as "the CEO of a profitable global enterprise" (Rodriguez 2010, x). "By calling herself a 'CEO' Macapagal-Arroyo represents herself not as a head of state but as an entrepreneur, the ideal neoliberal subject, who rationally maximizes her country's competitive advantage" (Rodriguez 2010, x). Like many other developing countries, the Philippines have complied with the mandates of the Washington Consensus and by the beginning of the twenty-first century the archipelago has seen the "hegemony of neoliberal globalization" (Rodriguez 2010, 147).

Neoliberalism and Mining in the Philippines

Prospects for the mining industry to the year 2000

Just as the Ramos administration acted to pursue neoliberal policies promoting foreign direct investment in general, it also acted to encourage foreign direct investment into mining in particular. In 1989, Guillermo Balce, the director of the Mines and Geosciences Bureau (MGB) of the Department of Environment and Natural Resources (DENR) and Michael Cabalda, the chief science research specialist of the MGB, participated in seminar jointly organized by the World Bank and the United Nations Department of Technical Cooperation and Development (Rovillos et al. 2003). This seminar was entitled "Prospects for the Mining Industry to the Year 2000" and it emphasized opening up more lands for mining in the developing world, allowing more foreign ownership of mining projects in developing countries and reducing the tax rates imposed on mining corporations by the governments of developing country (Rovillos et al. 2003). Through its participation in this seminar, the Philippine government began to become exposed to the thinking of the World Bank regarding mining as a development paradigm (Moody 2007). At this time there was no actual mining legislation in the archipelago and mining was being governed by two executive orders (Executive Order No. 211 and Executive Order No. 279), which were issued by President Aquino in 1987. These executive orders called for the formal enactment of mining legislation by the Philippine Congress and facilitated the authorization of mining projects by foreign corporations on the precondition that the project involved no more than 40 percent foreign ownership (Lopez 1992). The requirement for 60 percent Philippine ownership of mining projects in these executive orders was highly unpopular in the mining industry and the Chamber of Mines articulated a strong objection to this, arguing that many foreign mining companies were reluctant to invest in the Philippines without having full decision-making authority over their projects (Lopez 1992). During 1989, a working group of representatives from the Chamber of Mines of the Philippines[3] and the DENR began drafting a new mining code with assistance from the United Nations Development Programme, which was designed to allow 100 percent foreign ownership of mines in the islands (Lopez 1992). Then in 1992 the Philippine government, in conjunction with the Chamber of Mines, held seminars on the archipelago's mining investment opportunities in London, Manila and Vancouver (United States Bureau of Mines 1992). These seminars were designed to draw attention to the mineral resources of the islands so that, once a new mining code was passed, investors would be aware of the opportunities to avail them in the Philippines.

The Mining Act of 1995

On 6 March 1995 President Ramos signed into law the Philippine Mining Act of 1995 (Republic Act 7942), which had been Senate Bill No. 1639 and was sponsored by then senator Macapagal-Arroyo (Carreon 2009). The Mining Act was directed at adhering to the World Bank development model of a liberalized economy (Tauli-Corpus and Alcantara 2004). Clearly, it was designed to stimulate the economy by making the mining

---

3 The Chamber of Mines of the Philippines was organized in 1936 to serve as a lobby group for the mining industry. It lapsed into inactivity in the 1950s but was reconstituted in 1975 and has been a consistent and vigilant voice for the industry (Lopez 1992).

sector a driver of economic growth; as the deputy director of the MGB, Benjamin M. De Vera, wrote, "For a developing country like the Philippines, economic growth has always been considered as the primary objective of the government" (De Vera 1999, 259). The Mining Act contained a number of incentives to encourage mining in the archipelago such as a four-year income tax holiday, tax and duty-free capital equipment imports, value added tax exemptions, income tax deductions where operations are posting losses and accelerated depreciation. The statute also guaranteed the right of repatriation of the entire profits of the investment and guaranteed that no mining project would ever be expropriated. A guarantee of no expropriation is an extremely import enticement to mining investment (at least insofar as the guarantee itself can be trusted) because once developed a mine cannot be relocated. There is nothing more distressing to a mining company than spending several hundred million dollars in up-front costs on exploration and mine development only to see its mine expropriated by the host government. Perhaps the best examples of this were the 1971 nationalization of the mines owned by Anaconda Copper, an American mining company, in Chile (Holden et al. 2007). President Salvador Allende expropriated all of Anaconda's Chilean assets with no compensation whatsoever and Anaconda lost two-thirds of its production. This caused Anaconda to report a total loss for the year of USD 350 million and in today's prices this would be a loss for a single year of almost USD 1.9 billion.

However, perhaps the most significant aspect of the Mining Act of 1995 was the creation of two new types of production agreements governing the mineral deposit ownership requirements under which a foreign mining corporation could operate in the Philippines. These are the Mineral Production Sharing Agreement (MPSA) and the Financial Technical Assistance Agreement (FTAA). The MPSA is a production agreement that can last for up to twenty-five years, is approved by the DENR and requires that no more than 40 percent of the mineral project be owned by a foreign corporation. The FTAA, in contrast, is a production agreement that can last for up to twenty-five years, is approved by the president of the Philippines and allows 100 percent foreign ownership of the mine. The FTAA provisions and their facilitation of 100 percent foreign ownership were something the mining industry had persistently requested (*Mining Journal* 1993). When the Mining Act was passed, the industry felt it had finally gotten something it had always wanted; *Mining Journal* published an article entitled "Philippines Gets the Go-Signal," lauding its passage (*Mining Journal* 1995). From 1994–96 the number of foreign mining companies represented in the country increased by 400 percent (United States Geological Survey 1996). According to Snell (2004, 38), "Extractive industries from around the world staked out about

40 percent of the nation's land area in the Act's first two years." The United States Geological Survey (1997, x1) went so far as to call the Mining Act of 1995 "one of the most modern in Southeast Asia." By 2002, there were 43 FTAAs in place covering 2.2 million hectares (approximately 8 percent of the land area of the Philippines) in contrast to the 270,716 hectares, (approximately 0.9 percent of the land area of the Philippines) that were covered by MPSAs (Tujan and Guzman 2002).

A reinvigorated mining industry

Then, in 2004, two events occurred which solidified the role played by the Mining Act in attracting mining investment. First, on 16 January 2004, President Macapagal-Arroyo issued Executive Order 270 ordering the preparation of a Mineral Actions Plan directing the DENR to set guidelines and procedures for the simplification and streamlining of permitting and clearance systems for mines. In the Mineral Actions Plan, the first issue to be addressed was the "tedious permitting process" for mines and this was to be achieved by a simplification of procedures for the issuance of permits by the DENR. This executive order made it clear that the priority of the government would be the revitalization of the mining industry and that expediting mining projects was a matter of the highest priority (*Mining Journal* 2005). According to Carreon (2009, 105), Executive Order 270 was designed "to harmonize and synchronize the requirements and procedures in order to facilitate the inflows of investments into the mining industry." Clearly, the administration of Macapagal-Arroyo pursued a policy of the fullest possible liberalization of mining. Second, on 1 December 2004, the Supreme Court of the Philippines reversed its 12 January 2004 decision invalidating the FTAA provisions of the Mining Act (Ciencia 2006). This decision of the Supreme Court held that allowing foreign corporations to own 100 percent of a mine is inconsistent with the provisions of requirements of Section 2, Paragraph 4, Article XII of the 1987 Constitution. These provisions require that all natural resource extraction must be carried out by the state, carried out by a corporation whose capital is at least sixty percent Philippine or carried out by the state in an agreement with a foreign-owned corporation that provides financial or technical assistance (Ciencia 2006). In its 12 January 2004 decision, the Supreme Court held that 100 percent foreign ownership of a mine is inconsistent with a foreign-owned corporation providing financial or technical assistance (Ciencia 2006). On 1 December 2004, however, after 11 months of concerted pressure from pro-mining forces within the Macapagal-Arroyo administration (and after hearing numerous dire predictions of economic collapse), the Supreme Court reversed itself and held that 100 percent foreign ownership of a mine is consistent with a foreign-owned corporation providing financial or technical assistance and the FTAA provisions were upheld (Ciencia 2006). President Macapagal-Arroyo applauded this decision stating, "We are now poised for a strong investment take-off, based on a reinvigorated mining industry never seen in the recent past" (*Mining Journal* 2005, 21).

After the decision upholding the FTAA provisions of the Mining Act in December of 2004 Romulo L. Neri, secretary of the National Economic Development Authority (NEDA), presided over a mining conference in Manila in February of 2005 where the advantages of the archipelago's mineralization were touted to representatives of some of the world's largest mining companies (Neri 2005). Then, from 2005 to 2007, "a slew of foreign mining applications [were] being reported monthly by the government" (Ofreneo 2009, 202). Many of the world's largest mining companies expressed interest in investing in the Philippines with investment targets ranging from hundreds of millions to several billion dollars (Joint Foreign Chambers of the Philippines 2010). This commitment to mining continued after the departure of Macapagal-Arroyo from office in June of 2010 and in December of that year her successor, President Benigno Aquino III, declared his full support for the mining industry at the 57th Annual National Mine Safety Conference held in Baguio City (*Philippine Star* 2010). Clearly, by the end of the first decade of the twenty-first century the government of the Philippines had become bent upon a development strategy led by mineral resource extraction. As Oxfam Australia (2008, 9) wrote, "Since 2004, the Philippine Government has made a concerted effort to encourage the reinvigoration of the mining industry in the hope that mining will be a major driver of economic growth."[4] In the words of an unnamed exploration company president, "The Philippines has taken great strides in the last two years to attract investors through policy and promotion" (Fraser Institute 2008, 24). The success of these efforts to attract mining can be seen in Figure 2.1 where real foreign direct investment in mining over the 2004 to 2010 time period is depicted. Although mining investment does depict some volatility over this time period, (no doubt brought on by the controversy surrounding the constitutionality of the Mining Act during

---

4 The reader will note the frequent references to reports either authored or published by nongovernmental organizations (NGOs) in this book with Oxfam Australia (2008), Power (2002, 2008) and Ross (2001) being some of the salient examples. Much of the cutting edge research on mining as a development paradigm has been conducted by NGOs and the authors regard their reliance upon this literature as imminently reasonable. As Bebbington et al. (2008, 890) wrote, "Indeed, except as regards debates on mining and macroeconomic and political issues, the activist community has been well ahead of the scholarly community."

Figure 2.1. Mining investment in the Philippines, 2004–2010, in millions of 2004 constant dollars

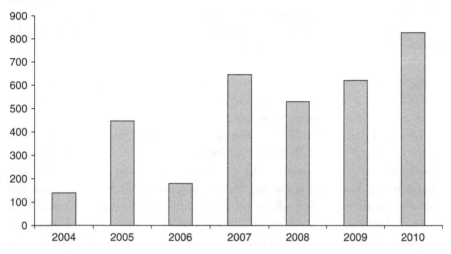

Source: Landingin (2008), Morales (2010), Olchondra (2011).

2004 and the effect of the global recession in late 2008) mining investment increased on average by 30 percent a year during this time period.[5]

Mining: A Leading Engine for Economic Growth

At the seventh Asia Pacific Mining Conference and Exhibition held in June 2007, President Macapagal-Arroyo declared that mining would "serve as a leading engine for Philippine economic growth, becoming a source of revenue and wealth to allow the government to seriously bring down the level of poverty in the country" (Ilagan 2009, 120). Clearly, the expectations placed upon the mining industry by the government are enormous. The value of the archipelago's minerals is estimated at USD 1.4 trillion which is equivalent to ten times the Philippines' gross domestic product and fourteen to seventeen times the country's foreign debt (Joint Foreign Chambers of the Philippines 2010). As Hatcher (2010, 1) wrote, "The considerable mineral endowment of the Philippines has, over the years, repeatedly been the mantle upon which indebted governments have stood to promise employment, education, health care, clean water and infrastructures." In terms of job creation, Romulo L. Neri indicated

---

5 Intertemporal averages such as this are calculated by taking the natural logarithm of the end value (USD 825,450,000), subtracting from that the natural logarithm of the beginning value (USD 139,500,000), dividing this difference by the number of years between the two observations (six) and multiplying by 100.

that mining can create 240,000 jobs (Neri 2005). Poverty eradication will occur because mining will result "in inducing greater economic growth, attracting more investments, creating more jobs and relieving poverty particularly in rural areas" (Neri 2005, 18). "Communities that play host to mining activities also benefit from improved infrastructures as well as other livelihood opportunities that these activities bring" (Neri 2005, 16). The government even sees mining as being something that can end the conflict between it and the NPA; according to Neri (2005, 21), "mining operations will also reduce insurgency in these remote areas as they generate jobs and reduce poverty."

In Figure 2.2 and Table 2.1, the locations of the major operating and proposed metallic mines in the Philippines are presented. Although mining is widespread in the archipelago (with mining projects being located on Luzon, Palawan, the Visayan Islands and Mindanao), the geological apriorism that metallic ore deposits generally occur in mountainous terrain has resulted in three distinct concentrations of mines: the Cordillera of northern Luzon, the northeast corner of Mindanao and the southeast corner of Mindanao.

Many of these mines are owned and operated to some extent by Australian and Canadian junior mining companies, often in conjunction with members of the oligarchy (Landingin 2008; Tujan and Guzman 2002). Many mining companies are headquartered in Australia and Canada due to their rich mining traditions and the ease with which exploration funds can be obtained on their stock exchanges (Australian Conservation Foundation 1994; Power 2002; Tujan and Guzman 2002). Australian mining companies have been aggressive investors in the Philippines since the mid-1980s, partly due to the proximity of the archipelago to Australia (Lopez 1992). Canadian mining companies are active worldwide accounting for 40 percent of all global exploration expenditure, particularly in the developing world (Canadian Dimension 2011). There are three reasons why Canadian mining companies are so active in the developing world: first, there are few high-grade ore deposits left in Canada; second, the Canadian government provides generous tax deductions and easy credit for Canadian mining operations in foreign countries; third, the Canadian government provides substantial diplomatic assistance[6] to Canadian mining companies in the event that they become

---

6 On 20 February 2007 the senior author of this book was preparing to interview Bishop Gabriel Peñate (of the Apostolic Vicariate of Izabal) in Puerto Barrios, Guatemala about the activities of Skye Resources, a Canadian mining company. Just as the interview was about to commence, the Canadian Ambassador to Guatemala telephoned Bishop Peñate and informed him that there was a group of Canadian antimining activists travelling around Guatemala and that Bishop Peñate should not talk to them. After a lengthy discussion between Bishop Peñate and the Canadian Ambassador the interview began with Bishop Peñate's disclosure of the ambassador's warning (Peñate 2007).

Figure 2.2. Locations of major operating and proposed metallic mines

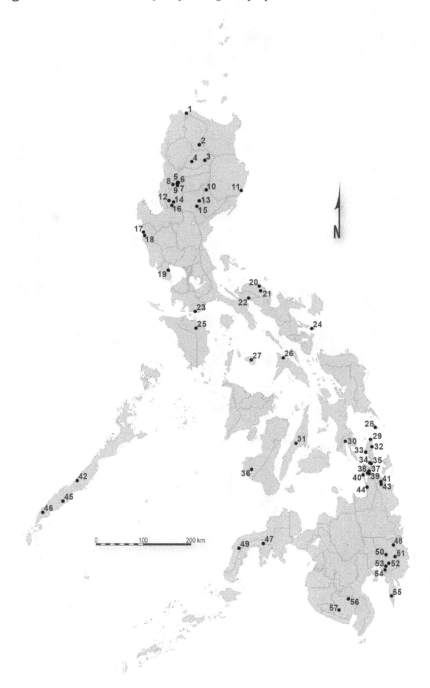

Source: Based on data from the Mines and Geosciences Bureau (2006, 2007a, 2007b, 2007c).

Table 2.1. Mine location information for the Philippines

Number (Figure 2.2)	Mineral	Name of project	Proponent of project	Province
1	Cu, Au	Claveria Copper-Gold Project	Oceana Gold	Cagayan
2	Cu, Au	Conner Copper-Gold Project	Cordillera Exploration Corporation	Apayao
3	Cu	Tabuk Copper Project	Wolfland Resources	Kalinga
4	Cu, Au	Batong-Buhay Copper-Gold Project	Batong-Buahy Gold Mines	Kalinga
5	Au	Victoria Gold Project	Lepanto Consolidated Mining	Benguet
6	Au	Far South East Gold Project	Lepanto Consolidated Mining	Benguet
7	Au	Teresa Gold Project	Lepanto Consolidated Mining	Benguet
8	Au	Gambang Gold Project	Oxiana Philippines Incorporated	Benguet
9	Au	Itogon Suyoc Gold Project	Itogon Suyoc Mines Incorporated	Benguet
10	Cu	Cordon Copper Gold Project	Vulcan Industrial Mining	Isabela
11	Ni	Dinapigue Nickel Project	Platinum Group Minerals	Isabela
12	Au	Camp 3 Gold Project	Northern Luzon Mining	Benguet
13	Au	Runruno Gold Project	Metex Mineral Resources	Nueva Vizcaya
14	Au	Acupan SSM Operations	Benguet Corporation	Benguet
15	Cu, Au	Didipio Copper-Gold Project	Oceana Gold	Nueva Vizcaya
16	Cu	Padcal Copper Expansion Project	Philex Mining Corporation	Benguet
17	Ni	Acoje PGE/Nickel Project	Crau Minerals	Zambales
18	Cr	Masinloc Chromite Project	Benguet Corporation	Zambales
19	Au	Bataan Gold Project	Balanga, Bataan Mineral Exploration	Bataan

	Project	Company	Commodity	Location
20	Paracale Gold Project	Johson Mining Corporation	Au	Camarines Norte
21	Labo Gold Project	El Dore Mining Corporation	Au	Camarines Norte
22	Del Gallengo Gold Project	Phelps Dodge Exploration	Au	Quezon
23	Lobo Gold Project	Mindoro Resources Limited	Au	Batangas
24	Rapu-Rapu Polymetallic Project	Lafayette Philippines Incorporated	Cu, Au, Ag, Zn	Albay
25	Mindoro Nickel Project	Crew Development Corporation	Ni	Mindoro Oriental
26	Masbate Gold Project	Filiminera Resources	Au	Masbate
27	Romblon Nickel Project	Pelican Resources	Ni	Romblon
28	Homonhon Chromite Project	Heritage Resources	Cr	Eastern Samar
29	Omasdang Chromite Project	CRAU Mineral Resources	Cr	Surigao del Norte
30	Sogod Copper-Gold Project	UP Mines	Cu, Au	Southern Leyte
31	Toldeo Copper Project	Atlas Consolidated Mining	Cu	Cebu
32	Vista Buena Mining II	Vista Buena Mining	Au, Cr, Ni	Surigao del Norte
33	Vista Buena Mining I	Vista Buena Mining	Au, Cr, Ni	Surigao del Norte
34	Nonoc Nickel Project	Pacific Nickel Philippines	Ni	Surigao del Norte
35	Sigbanog Nickel Project	Hinatuan Mining	Ni	Surigao del Norte

(Continued)

Table 2.1. Continued

Number (Figure 2.2)	Mineral	Name of project	Proponent of project	Province
36	Cu, Au	Colet Copper-Gold Project	Colet Mining	Negros Occidental
37	Cu, Au	Makalaya Copper-Gold Project	Manila Mining Corporation	Surigao del Norte
38	Cu	Boyongan Copper Project	Silangan Mindanao Mining	Surigao del Norte
39	Au	Siana Gold Project	JCG Resources	Surigao Del Norte
40	Au, Cu	Tapian San Francisco Project	Mindoro Resources	Surigao del Norte
41	Ni	Taganito Nickel Project	Taganito Mining Corporation	Surigao del Norte
42	Ni	Berong Nickel Project	Berong Nickel Corporation	Palawan
43	Ni	Adlay-Cagdianao–Tandawa Nickel Project	Case Mining Corporation	Surigao del Sur
44	Au, Cu, Ni, Co	La Fraternidad Project	San Roque Metals Inc	Agusan del Norte
45	Ni	Celestial Nickel Project	Toledo Mining Corporation	Palawan
46	Ni	Palawan Nickel Project	Coral Bay Mining	Palawan
47	Au, Ag, Cu, Pb	Balabag Project	TVI Pacific	Zamboanga Sibugay

48	Au, Ag	Diwalwal Direct State Development Project	National Resources Mining Development Corporation	Compostela Valley
49	Au	Canatuan Gold Project	TVI Pacific	Zamboanga del Norte
50	Au	Manat Gold Project	Indophil Resources	Compostela Valley
51	Cu	Tagpura Copper Project	Philco Mining	Compostela Valley
52	Cu	Amacan Copper Project	North Davao Mining Corporation	Compostela Valley
53	Au	Hijio Gold Project	North Davao Mining Corporation	Compostela Valley
54	Cu, Au	King King Copper-Gold Project	St. Augustine Gold and Copper Mining Limited	Compostela Valley
55	Ni	Pujada Nickel Project	Hallmark Mining Corporation	Davao Oriental
56	Cu, Au	Tampakan Copper Project	Sagittarius Mines Incorporated and Xstrata Copper	South Cotabato
57	Au	T'boli Gold Project	Cadan Resources	South Cotabato

Note: Ag = Silver, Au = Gold, Co = Cobalt, Cr = Chromium, Cu = Copper, Ni = Nickel, Pb = Lead, Pt = Platinum, Zn = Zinc.
Source: Mines and Geosciences Bureau (2006, 2007a, 2007b, 2007c).

involved in controversies pertaining to their operations in developing countries (Canadian Dimension 2011).

These companies are referred to as "junior" mining companies because they are small and have low amounts of capital; they exist by finding and developing ore deposits and then entering into partnerships with larger mining companies or by simply selling off their shares at wildly inflated prices. As F. Quimpo (2009, 79) wrote:

Junior mining companies which make up most of the mining companies in the world are relatively small corporations that engage in high-risk investments that may sometimes prove to be very profitable. They number about 3,067. They lack expertise and long-term capital to pursue mining projects after exploration, so they enter into partnerships with big corporations. They are usually financed by "smart money" on the stock exchanges of Canada and Australia, or get investments from pension, mutual and insurance funds. They are thus highly speculative and can rack up big share price gains without digging a hole.

Two mines which are particularly noteworthy for the purposes of this book are the Rapu-Rapu Polymetallic Project, on Rapu-Rapu Island in the Bicol Region of southeast Luzon and the Tampakan Project in the province of South Cotabato on the island of Mindanao (Figure 2.2 and Table 2.1). The Rapu-Rapu Polymetallic Project is a mine, originally developed by Lafayette Mining of Australia, which produces copper, gold, silver and zinc (Oxfam Australia 2008). This was the first new foreign-owned mine in the Philippines in over thirty-five years, the only zinc mine in the Philippines and the largest private venture in the province of Albay (Oxfam Australia 2008). The Tampakan Project is a copper and gold mine being developed by Xstrata of Switzerland and Sagittarius Mining Incorporated, a joint Australian–Philippine corporation (Xstrata 2008). With production scheduled to commence in 2016, Tampakan is believed to be one of the largest undeveloped copper and gold deposits in Southeast Asia (Estabillo 2009a; Xstrata 2008). Given the estimates that this mine will produce 340,000 tons a year of copper and 350,000 ounces a year of gold over an estimated twenty-year operational lifetime, the Tampakan Project is frequently referred to as a "world class deposit" (Estabillo 2009a; Xstrata 2008). Both of these mines embody the high hopes placed upon the mining industry by the Philippine government and both of these mines are also instrumental in highlighting the problems inherent in attempting to pursue a mining-based development paradigm in an area subject to natural hazards. However, before the discussion of how mining can adversely interact with natural hazards, attention must be directed towards mining's environmental effects, one of its most intractable aspects. Mining can generate a number of substantial environmental effects; these effects can occur over large distances, occur in perpetuity, can be extremely expensive to ameliorate and can have devastating consequences upon communities adjacent to mine locations. To count on an activity with such serious potential for environmental harm to act as a force capable of solving widespread poverty assumes – almost implicitly – that these environmental effects will either not occur or if they do occur, will be managed in such a way as to generate no significant adverse effects. Such assumptions, as the next chapter will show, range somewhere between optimistic to irresponsible.

Chapter Three

ENVIRONMENTAL
EFFECTS OF MINING

Mining: An Activity with Substantial Potential for Environmental Harm

The government of the Philippines is counting on the role mining can play as a source of economic development, but it must be acknowledged that mining is also an activity presenting a plethora of environmental, social and economic problems. As Pring et al. (1999, 45) acknowledged, "Few, if any, forms of economic development present the array of potential environmental, social and economic problems of the mining industry." Discovering, extracting and processing minerals is widely regarded as one of the most environmentally and socially disruptive activities in the world and environmental impacts can occur during exploration, mine development, mine operation and long after a mine has closed down (Bebbington et al. 2008). Modern large-scale mining can alter landscapes, water systems, economies and communities, often permanently (D'Esposito 2005).

Mining's Visual Impacts

The most obvious (and arguably least serious) impact of mining is its visual, or aesthetic, impact. "To conservationists," wrote Francaviglia (2004, 40), "mining landscapes seem a nightmarish expression of technology run amok." "Visually, mining produces some of the most dramatic landscapes on earth" (Francaviglia 2004, 42). In developed countries it is common for mining projects to be objected to because they are seen as industrial blight upon the landscape; surface disturbance, dust and vegetation removal all contribute to a general unsightliness of the landscape (Ripley et al. 1978). The proposed New World mine, in the American state of Montana, was objected to for a variety of reasons including the fact that it would be visible from within Yellowstone National Park and it would impair the backcountry experience availed by visitors to that park (Corkran 1996).

Impacts on Biodiversity

A more serious impact of mining is its impact upon biodiversity. According to the Mining, Minerals, and Sustainable Development report (2002)[1] mining almost always has an impact on biodiversity and leads to a reduction in the number of species present in the location of a mining project. The building of access roads into an undisturbed area can act as a vector for introduced species and lead to habitat fragmentation (Mining, Minerals, and Sustainable Development 2002). It is estimated that mining threatens approximately 40 percent of the world's undeveloped tracts of forest (Whitmore 2006). Aquatic species are often the most adversely effected by mining as the clearing of vegetation, shifting of large quantities of soil, extraction of large volumes of water and disposal of waste (either on land or in water bodies) can lead to soil erosion, sedimentation and the alteration of watercourses (Mining, Minerals, and Sustainable Development 2002). These environmental impacts can change the spawning grounds of fish and the habitats of other bottom-dwelling creatures.

In the Philippines, mining has clearly generated substantial impacts on the biodiversity of the archipelago, which has been described as "perhaps the greatest concentration of unique biological diversity of any country in the world" (Haribon Foundation and Birdlife International 2001, 17). It has been estimated that 19 percent of all important bird habitat has been directly lost to mining and that this figure rises when one takes into account the indirect effects upon habitat such as an influx of settlers who use mining roads (Haribon Foundation and Birdlife International 2001). Mining has also been a substantial factor leading to the loss of the unique pine forests of northern Luzon as these forests have been overexploited for mining timbers (IBON 2006a). Mining on Homonhon Island, off of the southeast coast of the island of Samar, has led to the near extinction on that island of the rare ironwood tree, a species once abundant in pre-Hispanic times (Santos and Lagos 2004). Even mine reclamation can have adverse impacts upon biodiversity. Well-intentioned efforts at revegetation of mine sites have served as a vector allowing the introduction of exotic species of vegetation and this can generate

---

1 The Mining, Minerals, and Sustainable Development report was a report funded by a number of large mining companies (such as Anglo American, BHP Billiton, Codelco, and Newmont Mining), NGOs (The World Conservation Union), government agencies (Environment Australia and Natural Resources Canada) and multilateral agencies (World Bank) to examine the environmental effects of mining and whether mining is something capable of generating sustainable development. Roger Moody, a noted antimining writer, referred to it as "probably the broadest-scoped critical examination of an industrial sector yet performed" (Moody 2007, 2), "possibly the most detailed dissection of a single industrial sector yet undertaken" (Moody 2007, 158) and "genuinely objective" (Moody 2007, 160).

many deleterious effects for endemic species living near the mine site (Mining, Minerals, and Sustainable Development 2002).

Acid Mine Drainage

Mining's most serious environmental effect

The most serious environmental effect associated with hardrock mining is the phenomena known as acid mine drainage. "Acid mine drainage is as tough an opponent as any industry has battled" (Stiller 2000, 97). Acid mine drainage is a function of the chemical composition of the ore bearing rocks. Often the rocks containing metals are found in what are known as sulfide ore deposits and one of the best examples of this is the mineral pyrite, which consists of iron and sulfur (Stiller 2000). When these sulfide ore deposits are exposed to water and oxygen, the pyrite will chemically react and create sulfuric acid (Smith 2005). Acid mine drainage is a natural phenomenon. In an undisturbed condition, rocks containing sulfides will generate acid at an imperceptibly slow rate (Stiller 2000). When, however, the rocks are broken during mining or crushed during processing, the acid generation will occur at a rate infinitely more rapid than ever found in nature (Smith 2005).

Measuring acid mine drainage

To measure acidity, chemists use the potential of hydrogen (pH) scale, which measures the hydrogen ion concentration in the solution of a substance; the lower the pH the more acidic and the higher the pH the more alkaline (Smith 2005). A pH of 7 is considered neutral (neither acidic nor alkaline) and is considered as safe for most living organisms (Regis 2008). The pH scale is logarithmic, so water with a pH of 3 is 10 times more acidic than water with a pH of 4 and 100 times more acidic than water with a pH of 5 (Smith 2005). Rainwater is slightly acidic (with a pH from 6.0 to 6.5) whereas seawater is slightly alkaline (with a pH from 7.8 to 8.2) (Regis 2008). Should a sulfide ore deposit become exposed to water and oxygen it will begin generating acid and this can cause the pH level of any receiving water to fall to a pH of around 4 (Stiller 2000). However, should certain species of bacteria (such as *Ferrobacillus ferrooxidans*, *Ferrobacillus sulfooxidans*, or *Thiobacillus thiooxidans*) be present, the generation of acid can accelerate substantially and pH may decline below 3 (Smith 2005). The classic telltale sign of acid mine drainage is a reddish-orange color in an affected water body; this indicates a rapid rate of pyritic iron oxidization (Stiller 2000). Accordingly, acid mine drainage poses a very serious challenge for aquatic biota as most fish populations cannot breed below a pH of 5, and some populations even die out at a pH of 6 (Ripley et al. 1996).

The mobilization of heavy metals

The acidity imparted into a water body by acid mine drainage is, however, only a starting point. As acid mine drainage continues, heavy metals (called "heavy" metals because they have a high specific gravity) can be mobilized from the ore-bearing rocks and become present in mining affected waters (Regis 2008; Smith 2005; Stiller 2000; Ripley et al. 1996). These heavy metals, such as arsenic, cadmium, chromium, copper, lead, mercury and zinc, can have serious consequences on living organisms and their effects on human health are presented in Table 3.1.

What makes these heavy metals so serious is the potential they contain for biomagnification in the food chain (Regis 2008). One of the most well-known examples of biomagnification is mercury, which may enter aquatic food systems where it is converted into the poisonous neurotoxin methylmercury. Since this compound is produced at a rate faster than that which fish can degrade it, it will accumulate in fish and then when those fish are eaten by humans, will be conveyed into the bodies of those who eat the fish. Mercury has extremely serious effects upon the human nervous system and can cause brain damage. The most infamous example of this being the mercury poisoning suffered by the residents of Minamata, Japan who were exposed to mercury emitted in wastewater.

Acid mine drainage: Difficult to predict and prevent, impossible to stop

Acid mine drainage is difficult to predict, difficult to prevent, slow to develop and impossible to stop once underway. Sulfide ore bodies tend to be heterogeneous in their geochemical content (Stiller 2000). This means that they can consist of pockets of acid generating material mixed among ore that does not contain the potential to generate acid. To find these pockets of acid generating material, extensive sampling must be undertaken before mining begins and this sampling (akin to trying to estimate the contents of a library by withdrawing several dozen books) can be difficult to do with the necessary degree of accuracy (Stiller 2000). Also, the more thorough the sampling becomes, the more time will be spent on sampling and the greater the expense of sampling will become. Mining companies are in business to extract minerals, not to produce voluminous accounts of the geochemical conditions present at their mines, and ultimately, cost conditions limit the amount of sampling that can be undertaken. There have been numerous instances where acid mine drainage has developed even though it was not predicted to occur (National Research Council 1999). Preventing acid mine drainage by avoiding acid generating materials is, in the words of Stiller (2000, 97), "an unimaginable task." In most instances it is nearly impossible to segregate one set of metal sulfides from another and selectively "mining particular zones to avoid pyrite is rarely possible, practical, or effective" (Stiller 2000, 97). Acid mine drainage may take years to develop after mining has started and the generation process is not well enough understood to determine how high the concentrations of acid and metals will reach (National Research Council 1999). Lastly, once acid mine drainage has started it can be ameliorated at tremendous expense by running the acidic waters through limestone or some other type of buffering agent, but it can never be stopped (Stiller 2000).[2] In Europe, there are lead mines that were developed during the Roman Empire; although these mines would be considered small-scale mines by today's standards they are still producing acid mine drainage today, over 2,000 years later (National Research Council 1999). Once acid mine drainage starts, the proverbial genie is allowed to escape from the proverbial bottle and the process can never be stopped; a geochemical process will now be underway and it will operate on a geological time scale.

In Wisconsin, in the United States, concerns about acid mine drainage were so serious that they led to the passage of the sulfide mining moratorium by the state legislature in 1998 (Gedicks 2001). This moratorium prohibits a mining company from opening a mine in a sulfide ore body unless that company can demonstrate that it has operated a mine in a sulfide ore body for a minimum of 10 years with no acid mine drainage (Gedicks 2001). The moratorium also requires any mining company proposing the development of a mine in a sulfide ore body to prove that it has also maintained a mine developed in such an ore body for a minimum of 10 years after closure without acid mine drainage (Gedicks 2001).

Subaqueous Tailings Disposal: The Solution to Acid Mine Drainage?

Since the 1980s, the mining industry has become more cognizant of the seriousness of acid mine drainage (Ripley et al. 1996). The principal method of dealing with this problem is a process known as subaqueous tailings disposal wherein finely grounded mine wastes, or tailings, are deposited under deep water, thus placing them in an anaerobic environment and denying them the oxygen required for the process (Ripley et al. 1996). Usually these wastes are submerged behind a "tailings dam" in a "tailings pond" (see Figure 3.1).

Tailings dam failures

The importance of subaqueous tailings disposal as a method of managing acid mine drainage leads to the environmental risks posed by tailings dams

---

2 See Chapter Seven for a discussion of the costs involved in mine remediation.

Table 3.1. Characteristics and effects upon human health of select heavy metals

Heavy metal	General characteristics	Effect on human health
Antimony	Similar properties to arsenic, but much less toxic. Often found in gold bearing ore bodies.	Long-term inhalation may cause lung disease, skin, eye and throat irritation, dizziness and headaches, nausea and vomiting, diarrhea; stomach cramps, insomnia, and an inability to smell properly.
Arsenic	Arsenic can be acutely toxic in low concentrations. Arsenite is more toxic than arsenate. Often associated with ores containing gold, copper, and zinc. Arsenic trioxide is produced during gold roasting.	High doses lead to muscle spasms, nausea, vomiting, abdominal pain, diarrhea and death. Low doses over many years may result in skin, lung or bladder cancer; peripheral nervous system disorders, and dermatitis.
Beryllium	Toxic. Main hazard is the inhalation of dust and its salts.	The symptoms of beryllium poisoning (weight loss, weakness, chest pain, coughing, eye irritation) may be delayed from 5 to 10 years. Beryllium is a potential carcinogen and can cause lung disease.
Cadmium	Highly toxic. Often found in ore bodies containing silver and zinc.	Cadmium is a known carcinogen. Cadmium accumulates in the liver and kidneys. Chronic exposure can cause renal damage, anemia, osteoporosis and osteomalacia.
Chromium	Chromic acids and salts are toxic. Chromium is an essential trace element, which can be toxic at higher levels.	Toxic symptoms of chromium include dermatitis, irritated eyes, scouring, dehydration, and skin allergies. Increased human hypertension, diabetes and death have been linked to chromium concentrations in stream sediments. Chromium tends to accumulate in the kidneys, bone, spleen, lungs, and liver. Chromic acid and chromate are potential carcinogens that may also cause liver, kidney and respiratory damage as well as skin irritation.

Copper	Often found in ore bodies that contain silver and zinc.	May cause eye and nose irritation, dermatitis, anemia, gastric ulcers, renal damage, and hemolysis. Greater risk for those with Wilson's disease.
Lead	Found in most ore bodies with copper, silver, and zinc.	Affects the digestive, blood, and nervous systems. May lead to spinal deformities, kidney dysfunction and hyperactivity. Possible carcinogen.
Manganese	Main concern is the generation of fugitive dust emissions.	Inhalation of excessive manganese dust is toxic. Affects nervous system. Symptoms include insomnia, mental confusion, dry throat, coughing, chest tightness, respiratory difficulty, flu-like fever, low back pain, vomiting, malaise, fatigue and kidney damage.
Mercury	Mercury is toxic if ingested or if inhaled in vapor form. Methyl mercury is one of the most hazardous environmental pollutants known.	Affects nervous systems, kidneys, eyes, and skin. Mercury poisoning can lead to brain and spinal cord damage, and is a possible carcinogen.
Nickel	Nickel carbonyl is highly toxic.	Nickel is a potential occupational carcinogen which accumulates in the kidneys. The central nervous system can be severely affected. Nickel carbonyl is carcinogenic.
Selenium	Selenium is an essential element, which can also be toxic.	Excessive amounts of selenium can cause tooth decay. Selenium is a possible carcinogen.
Zinc	Zinc is often found in ore bodies containing copper, lead and silver.	Zinc poisoning may lead to chills, muscle aches, nausea, fever, coughing, weakness, headache, blurred vision, lower back pain, vomiting, fatigue and breathing difficulty.

Source: Moodie (2001).

Figure 3.1. Tailings pond, Manila Mining Corporation's Placer Mine

themselves. In recent years, modern mining technologies have facilitated profitable extraction of metals from low-grade ore bodies and in Table 3.2 the average grades of major minerals are presented.

With such low grades of ore being mined, there will be substantial amounts of waste being generated. In modern copper mining, an ore grade of 0.91 percent will mean that 99.09 percent of all material mined will be waste rock and in modern gold mining an ore grade of 0.0003 percent will mean that 99.9997 percent of all

Table 3.2. Average grades of major minerals

Mineral	Average grade (%)
Chromium	30.0
Lead	2.5
Nickel	2.5
Copper	0.91
Gold	0.0003

Source: Warhurst (1994).

material mined will be waste rock (Francaviglia 2004). Most of the wastes produced by mining are the tailings and these usually contain residual chemicals such as cyanide and elevated levels of heavy metals (Mining, Minerals, and Sustainable Development 2002). For tailings dams to be an effective method of managing waste, water must always be over the material and this necessitates perpetual maintenance (National Research Council 1999). In the words of Mining, Minerals, and Sustainable Development (2002, 241), "Tailings storage facilities require not just good design but also close, consistent, routine attention over a long period." Should one of these tailings dams begin to leak or abruptly break, there will be a release of contaminants into the environment – an event that will significantly worsen environmental conditions (Mara and Vlad 2009). For the mining industry, this is a serious concern and the most widely reported environmental incidents "have been of breaks and spills from tailings dams and the discharge of tailings into rivers and waterways" (Burke 2006, 227). "There are few occurrences more dramatic, or traumatic, than the collapse of a mine tailings dam" (Moody 2007, 144). Globally, the greatest single concern to the mining industry is tailings dam failures and "major accidents seem to occur on average once a year" (Mining, Minerals, and Sustainable Development 2002, 240). From 1995–2006 there were at least twenty-eight tailings dam failures worldwide (Moody 2007). There is also evidence that the number of tailings dam failures is increasing over time. During the time period 1969–79 there were 10 tailings dam failures; during the time period 1989–99 there were 21 tailings dam failures (Stenson 2006). This increase in the number of tailings dam failures can be attributed to an increase in the amount of low-grade ore deposits being extracted by new technologies (Whitmore 2006). It must, however, be stressed that there is no complete worldwide database of all historical tailings dam failures and the majority of tailings dam failures – especially in developing countries – remain unreported (Rico et al. 2008a).

Selected tailings dam failures

Some selected tailings dam failures are detailed below. In Montana in 1975, heavy spring rains combined with melting snow caused the tailings dam at Anaconda Copper's Mike Horse Mine to spill approximately 100,000 tons of pyritic metal-bearing tailings into the upper Blackfoot River (Stiller 2000). In 1991, a beaver dam raised water levels in a lake that overflowed into the tailings dam at the Matachewan gold mine in Ontario, Canada; this caused the tailings dam to break, releasing tailings into the Montreal River (Ripley et al. 1996). In 1995, a tailings impoundment failure at the Omai gold mine in Guyana caused more than 4 billion cubic liters of containment wastewater to be spilled into the Essequibo River over a period of four days (Gedicks 2001). A study by the Pan American Health Organization showed that all aquatic life in the 4-kilometer creek that runs from the mine to the Essequibo had been killed (Gedicks 2001). In 1998, a tailings dam burst at the Los Frailes mine in Spain; spilling five million cubic meters of toxic waste into a river; the spill flooded between 5,000 and 7,000 hectares of farmland, destroyed bird habitats and killed 26 metric tons of fish (World Wildlife Fund and International Union for the Conservation of Nature 1999). The losses to fishing, agriculture and tourism were estimated at between USD 100 million to USD 200 million (World Wildlife Fund and International Union for the Conservation of Nature 1999). In 2000, at the Baia Mare gold mine in Romania, a tailings spill released 100,000 cubic meters of tailings, which resulted in a contamination of the Danube River used for drinking water by more than 2 million people (Moody 2007).

The Marcopper tailings spill on the island of Marinduque

Perhaps the most notorious example of a tailings impoundment failure occurred in the Philippines itself in 1996 at the Marcopper Mine on the island of Marinduque (Figure 3.2). The Marcopper Mine was located in the north central highlands of the island of Marinduque and was owned by the Marcopper Mining Corporation which was, in turn, owned (39.9 percent) by Canada's Placer Development and (60.1 percent) by the government of the Philippines (Futures Group International

Figure 3.2. The island of Marinduque

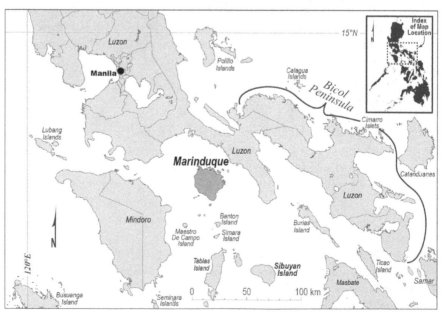

2004). Although the mine was jointly owned by Placer Development and the government, it was "under design and management control of Placer Development" (Futures Group International 2004, 1–2). Marcopper was the first major mine in the Philippines to be developed and brought into full-scale production as an open-pit mine (Lopez 1992). Copper began to be extracted from the Tapian pit in 1969 and by 1970 Marcopper was second only to the Toledo Mine on the island of Cebu in terms of copper production (Lopez 1992). Ore was extracted from this pit until 1991 when production switched to the San Antonio pit several kilometers to the north (Plumlee et al. 2000). In 1991, the mined-out Tapian pit had its dewatering drain plugged with concrete and it began to be used as an impoundment for the tailings from the newer San Antonio pit (Plumlee et al. 2000). By December 1995 a total of 32,476,841 metric tons of tailings were impounded in the Tapian pit (Mines and Geosciences Bureau 2004).

On 25 August 1995, problems began at the mine when excessive seepage began to occur from the plug at the bottom of the Tapian pit (Futures Group International 2004). This seepage continued unabated until 24 March 1996 when the plug failed completely and there was an abrupt release of tailings into the Boac River (Plumlee et al. 2000). The actual amount of tailings that was released is a matter of controversy; low-end estimates put the amount at 1.6 million cubic meters (Mines and Geosciences Bureau 2004) while high-end estimates put the amount at 3 million cubic meters (Plumlee et al. 2000). Regardless of the precise amount of tailings that were released, its effects were dramatic: five villages had to be immediately evacuated and an estimated 20,000 people were displaced (Stark et al. 2006). In April 1996, a UN team declared the river "biologically dead" (Mining, Minerals, and Sustainable Development 2002, 208). When the investigative team sent by the United States Geological Survey and the United States Armed Forces Institute of Pathology visited Marinduque in 2000 (four years after the tailings release) they reported that there were "still extensive tailings deposits visible in many places along the Boac River streambed" (Plumlee et al. 2000, 22). Their conclusion was that "the mining-environmental impacts on some parts of Marinduque have been substantial and pose significant long-term challenges for remediation, both from a technological and monetary standpoint" (Plumlee et al. 2000, 41). When another investigative team sent by the Futures Group International and the United States Geological Survey visited Marinduque in 2003, (seven years after the tailings release) they reported that depressed pH levels were detected at least 10 kilometers downstream of the mine and that "acid rock drainage from Marcopper is clearly having significant, far reaching and adverse impacts on surface water quality" (Futures Group International 2004, 5–66). Furthermore, "several heavy metals were at levels considered to be acutely or chronically toxic to aquatic life" (Futures Group International 2004, ES–5).

While the Philippine government prefers to refer to the tailings spill as "an incident" (Cabalda 2004, interview); others have referred to it as "the infamous tailings spill incident" (Rovillos et al. 2003, 202) or as a "disaster" (Gedicks 2001, 26; Lansang 2011, 129; Tujan 2001, 154; Tujan and Guzman 2002, 204). The tailings release generated a substantial amount of concern among the Philippine people about the environmental effects of mining. According to Tujan (2001, 154), "the Marcopper accident shocked and traumatized the Philippine nation." The incident caused the nation to give a "collective gasp of disbelief" (IBON 2006a, xviii). Bello et al. (2009, 224) described the spill as, "the biggest industrial accident in the country's history." To Hatcher (2010, 9), the spill has come "to be known as the worst environmental incident ever sustained in the Philippines." In the words of Stark et al. (2006, 5) the spill "has taken on mythic proportions in the Philippines." Ofreneo (2009, 202) wrote that the spill caused "not only untold damage to the ecosystem but also public alarm over the environmental impact of mining." The Foundation for Environmental Security and Sustainability 2007, 22) declared that whenever mining in the Philippines is discussed, the one incident that "is invariably mentioned by citizens and government officials alike is the Marcopper mine." Even Chris Hinde, the editorial director of *Mining Journal* (a mining industry publication), went so far as to call the Marcopper tailings spill "an environmental disaster" (Hinde 2004, 1). Michael Cabalda, chief science research specialist of the MGB, grudgingly acknowledged that whenever mining is discussed, "it is always Marcopper that is talked about" (Cabalda 2004, interview).

Clearly, a tailings dam failure leads to more than just a "mess" or an "eye sore," accompanied by a temporary halt in production at the mine with some environmental harm that is of a merely transitory nature. A tailings dam failure can lead to long lasting environmental degradation permanently degrading the quality of the environment depended upon by a local community and leaving them struggling to cope with its after effects for years into the future.

Chemical Spills from Mining Operations

Cyanide: A chemical agent used in modern mining

The principal development in mining technology facilitating the profitable extraction of metals (particularly gold and silver) from low-grade ore deposits has been the development of cyanide leaching (National Research Council 1999). In cyanide, vat leaching crushed ore is placed inside a steel vat containing a solution of sodium cyanide; in cyanide heap leaching crushed ore is placed on a polyurethane pad and then soaked with a solution of sodium cyanide (Holden et al. 2007). The difference between vat leaching and heap leaching is the grade of the ore; the former is used with relatively higher-grade ores and the latter is used with relatively lower-grade ores because a larger quantity of ore can be mixed with a solution of sodium cyanide when it is placed on a heap rather than being contained within a vat (Holden et al. 2007). In either case, the advantage of using cyanide leaching is that the microscopic particles of gold and silver chemically bond with the sodium cyanide, thus facilitating the extraction of low-grade ore deposits (National Research Council 1999).

Cyanide leaching is one of the most controversial aspects of modern hardrock mining due to the extreme toxicity of cyanide, "one teaspoon of two percent cyanide solution can kill a human" (Moran 1998, 1). From 1991 to 2006, there were at least nine major accidents involving cyanide at mining projects (Stenson 2006). Spills have also occurred while transporting cyanide to mines. In 1998 a truck transporting cyanide to the Kumtor gold mine in Kyrgyzstan fell off a bridge and spilled approximately two tons of sodium cyanide into a river; in the aftermath of the spill hundreds of people were treated at local hospitals, and at least one fatality was reported (Moran 1998). In 2000, a 1-ton bale of concentrated sodium cyanide fell from a sling underneath a helicopter en route to the Tolukuma gold mine in Papua New Guinea (Mineral Policy Institute 2000). The bale of cyanide fell through the canopy of the rainforest and landed only 20 meters from a steam. Tests showed cyanide contamination of 2,800 parts per million on the soil surface at the spill site with cyanide contamination of 5.4 parts per million 30 meters downstream from the impact site. For perspective, the United States' limit of cyanide for drinking water is 0.2 parts per million (United States Environmental Protection Agency 2009).

Fish are extraordinarily vulnerable to cyanide. Birds and mammals will be killed by cyanide concentrations in the milligram per liter range; fish, however, will be killed by cyanide concentrations in the microgram per liter range (Moran 1998). Even when cyanide breaks down naturally in the environment, it still generates various compounds as by-products. These by-products may persist naturally in the environment and there is evidence that some forms of these compounds can accumulate in plant and fish tissues (Moran 1998). There is also much uncertainty about the toxicity of cyanide and it is unclear how quickly, or slowly, cyanide will breakdown given a range of temperature and other environmental conditions; this may be the most problematic aspect of its use in hardrock mining (Moran 1998). In the state of Montana, a series of cyanide leaks from mining operations led to a citizen initiative outlawing the use of both cyanide vat and cyanide heap leaching (Holden et al. 2007).

Mercury: A by-product of mining

Toxic chemical spills have also occurred while transporting by-product chemicals such as mercury from mines. Liquid mercury is often produced as a by-product of hardrock mining and in 2000 in Peru, 151 kilograms of liquid mercury was spilled over a 45-kilometer stretch of road in the village of Choropampa while being transported from a gold mine (Ingelson et al. 2006). After the mercury was spilled, curious local residents (including children) gathered mercury with pieces of paper or their bare hands and stored it in their homes; some even thought the mercury was a medicinal elixir and tasted it. This resulted in more than two hundred people being hospitalized due to mercury poisoning.

Mining's Impacts on Water Resources

Even if a mine generates no acid mine drainage, has an impenetrable tailings dam and spills no chemicals, mining may have serious impacts upon water resources in terms of its impacts upon water availability. "It is well known that mining activities can lead to modifications of hydro-geological conditions" (Mara and Vlad 2009, 972). Mines are heavy water users, as Shomaker (2005, 128) wrote:

> Mining enterprises always require water. The uses vary widely: dust suppression, milling, and processing, conveyance of tailings from mills, recovery of metals by leaching, dewatering of underground workings and reclamation of mined lands. Apart from simple dewatering, most of these uses can lead to relatively high depletion – most of the water is lost to evaporation, rather than to the surface water or ground water system.

The most obvious effect of mining upon water availability is through the process of mine dewatering. If a mine goes below the water table it will flood with water unless water is continually pumped out of the mine (Smith 2005). In the case of open-pit mines, a lake will form within the pit where the ground water reestablishes itself (Figure 3.3); these lakes are referred to a "pit lakes" (Smith 2005).

To prevent pit lakes from forming many mines engage in extensive mine-pit dewatering programs, continually pumping groundwater out of the pit. While this may prevent the pit from flooding, it "often involves the dewatering

Figure 3.3. Pit Lake, Manila Mining Corporation's Placer Mine

of large areas, which can create long-term ground water deficits and affect surface water flows as well as seeps and springs" (Kuipers 2005, 48). As Ali (2003, 15) wrote:

> Water within a mine has been traditionally considered a hindrance to mining; hence, drainage programs from the mining site have caused major disruptions in groundwater regimes. The direction of groundwater movements may easily change due to mining, thus leading to disruptions in recharge regimes and the drying up of certain springs.

If there is one isolated mine in an area, groundwater withdrawal at this single mine has the potential to create a deep cone of depression in the local aquifer near that mine. If successively more water is removed from this mine, over time the cone may join those created by neighboring mines and this will lower the regional water table. This may lead to a reduction in flows in streams and springs some distance from the mine. Consequently, an important concern in mining is what the National Research Council (1999, 158) called:

> The potential cumulative effects of dewatering wells and groundwater wells for processing associated with several neighboring mines. The

cones of depression in the deep aquifer resulting from groundwater withdrawal may coalesce and affect regional spring and stream flows that are dependent on the aquifer.

The impacts of mining upon hydrologic resources are extremely important because water is a resource universally required by all people, regardless of their social class or economic sector. Given this acknowledgement of the impacts upon society stemming from mining's impacts upon the biophysical environment, attention now turns towards the social impacts of mining.

Mining's Impacts upon the Social Environment

Lastly, just as mining may have serious impacts upon the biophysical environment, it may also have correspondingly serious impacts upon the social environment. According to Anderson (1998, 16) these can include: "Prostitution, alcoholism, increased domestic violence, organized crime, cultural disruption, sexually transmitted diseases – all are on the list of potential social effects which can plague a community when a mine is in the construction phase." A sudden influx of new people and activities in an area can also serve as a disease vector having hugely detrimental effects upon human health (Mining, Minerals, and Sustainable Development 2002).

One of the most controversial aspects of mining is the deployment of security forces by mining companies to protect their operations (Holden and Jacobson 2007b, 2011; Holden et al. 2011). Mines involve substantial up-front expenditures and (particularly in the case of precious metals mines) involve the extraction of highly valuable materials. Consequently, mines are highly vulnerable targets for criminals or armed antistate groups.[3] Consequently, many mining companies hire armed security personnel or are protected by the police or military forces of the host government. Frequently, however, allegations arise that these "security forces" are simply there to intimidate those who oppose mining into being quiet and to refrain from engaging in antimining activism (Holden and Jacobson 2007b, 2011; Holden et al. 2011). Globally, there have been numerous reports of antimining activists being killed by security forces set up to protect mining projects (Moody 2007). As Pegg (2006, 385) wrote, "Resource extraction firms have been complicit in repeated human rights violations involving security forces."

---

3 The problems posed by the intersection of large-scale corporate mining amid the presence of the armed insurgent groups of the Philippines is discussed in Chapter Seven.

Mining: Clearly an Activity with Substantial Potential for Environmental Harm

Hardrock mining is an activity with substantial potential for environmental harm. Mining can impact biodiversity, lead to the generation of acid mine drainage, occasion the release of hazardous materials, result in chemical spills, reduce water resources and have pronounced social impacts. However, as attention turns to the examination of how the natural hazards present in the Philippines can adversely interact with mining, two of these impacts become salient: tailings spills and impacts upon ground water resources.

Modern mining methods can facilitate the profitable extraction of minerals from low grade ore deposits. However, these mining methods generate substantial amounts of waste in the form of tailings and often these tailings are highly acidic and contain cyanide. To prevent these tailings from being released into the environment and to stop the generation of acid mine drainage, tailings are submerged in water and kept in tailings ponds behind tailings dams. Since the geochemical process of acid generation is something operating on a perpetual time scale, these tailings dams will require perpetual attention. In this regard, they will be akin to the facilities used to store radioactive waste; they must always be watched and they must always be monitored. Should something occur that causes a tailings dam to fail, the consequences will be sudden, abrupt and catastrophic: a massive quantity of highly toxic material will immediately be released into the environment. Consequently, any potential threat to the stability of a tailings dam is an extremely serious concern because this will be something quite likely to exacerbate mining's environmental effects. While tailings dams are supposed to last forever, past experience shows that tailings spills pose a serious environmental threat staying behind long after any mine closes (Rico et al. 2008b). After the 1975 Mike Horse Mine tailings dam collapse, an ecologist employed by the Montana Department of Fish and Game "compared tailings impoundments to 'time bombs that are scattered wherever there has been mining'" (Stiller 2000, 79). To this ecologist, these tailings dams are just sitting there (ticking away) waiting for some event to cause them to release their tailings (Stiller 2000). The Philippines, with its vulnerability to typhoons, earthquakes, tsunamis and volcanoes, is a place where circumstances are present that are capable of causing a sudden and abrupt release of tailings. In many parts of the archipelago, the stability of tailings dams cannot be taken for granted.

Mining is also a heavy user of water. Mines often go below the water table and to prevent pit lakes from forming, the mines are dewatered. If a substantial amount of dewatering occurs at a mine, a cone of depression in the aquifer will develop around that mine. Should this cone of depression in the aquifer become large enough, it will join those created by other mines and a widespread drawing down of the water table will occur. Should there be a large concentration of mines in an area that are all engaged in mine-pit dewatering, ground water resources may become compromised over a large area. Should some event occur which abruptly limits precipitation and generates a drought, these ground water resources will become essential for the survival of nearby communities. By denying these communities ground water resources, mining may pose a serious challenge to them. Many parts of the Philippines are vulnerable to such a precipitation-reducing event capable of generating severe droughts: the El Niño Southern Oscillation (ENSO). Attention now turns to the natural hazards prevalent in the archipelago. What natural hazards are encountered by the residents of the archipelago? What are the causes of these hazards? Is the magnitude of these hazards constant or, given global climate change, will some (or even all) of these hazards worsen in magnitude and frequency? How can these hazards adversely interact with mining? Is it possible that these hazards may interact with each other, thus generating effects substantially greater than those occurring if they occurred in isolation? Is it possible that these hazards may also interact with other types of anthropogenic environmental degradation occurring in the archipelago and have an even more pronounced impact upon mining? These are the questions the next chapter will address.

Chapter Four

MINING AMID NATURAL HAZARDS

The Philippines: Spaces of Hazard

Natural hazards are those atmospheric, hydrologic, geologic and other naturally occurring physical phenomena that have the potential to harm humans (Punongbayan 1994). An extensive body of literature exists documenting the vulnerability of the Philippines to natural hazards (Bankoff 1999, 2003a, 2003b; Bankoff and Hilhorst 2009; Delica 1993; Holden 2011; Luna 2001; Yumul et al. 2011).

Typhoons: One of the World's Most Powerful Atmospheric Phenomena

Typhoons: Tropical cyclones in the Western Pacific

Typhoon, originating from the Chinese *tai* (strong) and *fung* (wind), is the term used to describe a tropical cyclone in the Western Pacific Ocean (Bankoff 2003a).[1] A typhoon is one of the world's most powerful atmospheric phenomena with a fully developed typhoon releasing the energy equivalent to many Hiroshima-sized atomic bombs (Wisner et al. 2004). Typhoons develop in the northern hemisphere during the months of July to November in an area just north of the equator (Bankoff 2003a; Wisner et al. 2004). They develop when strong clusters of thunderstorms drift over warm ocean waters having a temperature of at least 26.5 degrees Celsius. Warm air from the thunderstorms combines with the warm air from the ocean surface and begins rising; as this air rises there will be a reduction in air pressure on the surface of the ocean. As these clusters of thunderstorms consolidate into one large storm, trade winds blowing in opposite directions will cause the storm to begin spinning in a counter clockwise direction, while rising warm air causes air pressure to decrease at higher altitudes (Gonzalez 1994; Van Aalst 2006). Eventually, the

---

1 In the western hemisphere, tropical cyclones are called hurricanes from the Spanish *huracanes* (Bankoff 2003a).

storm will have a low-pressure center (the eye) with no clouds and no winds while winds in the outer part of the storm can exceed 118 kilometers per hour (Gonzalez 1994). Winds this strong are certainly capable of generating extremely serious damage.

The four characteristics of typhoons

All typhoons have four crucial characteristics: strong winds, heavy rains, storm surge and low air pressure (Bankoff 2003a; Gonzalez 1994; Wisner et al. 2004). The air pressure reduction associated with a typhoon is one of its most important aspects as this can cause the sea level to rise by as much as 1 centimeter for every 1 millibar reduction in air pressure (Wisner et al. 2004) and there have been documented instances during typhoons where sea levels have risen by 1.5 meters as a result of air pressure reductions (Wang et al. 2005). This means that at the same time that a typhoon (with its heavy rains and strong winds) surges ashore onto a piece of land, the sea level around that piece of land will be higher due to the reduction in the atmospheric pressure (Gonzalez 1994). Although typhoons rapidly lose power as they move inland, they are capable of causing massive amounts of damage to coastal areas and are extremely difficult to track accurately and can move in an unpredictable manner (Gonzalez 1994; Wisner et al. 2004).

Vulnerability of the Philippines to typhoons

Much of the Philippines are at risk from typhoons and each year about twenty of them, equivalent to 25 percent of the total number of such events in the world, occur in Philippine coastal waters (Wisner et al. 2004; Yumul et al. 2011). Approximately 95 percent of these typhoons originate in the Pacific Ocean – south and east of the archipelago – between the months of July to November and travel in a northwesterly direction mainly affecting the eastern half of the country (Bankoff 2003a; Yumul et al. 2011). The most heavily affected portions of the Philippines are Northern Luzon, the Bicol Region and Samar (Bankoff 2003a, 2003b). Mindoro, Panay and Leyte are at moderate risk to typhoons while Mindanao and Palawan are islands with a very low risk to typhoons (Bankoff 2003a, 2003b). Although the moisture provided by these storms has somewhat of a positive effect, (providing between 38 to 47 percent of the archipelago's average annual rainfall) overall their effects are profoundly negative in that they set off landslides, cause severe and recurrent flooding of lowland areas and are responsible for more loss of life and property than any other natural hazard (Bankoff 2003a, 2003b).

Mining and typhoons

The overlap of mining projects with the risk of typhoons is provided in Figure 4.1, which shows how one-third of all mining projects in the Philippines are in areas that are at high risk of being impacted by a typhoon while 7 percent of all mining projects are in areas that are at very high risk of being impacted by a typhoon. Much of the overlap between typhoon vulnerability and mining projects occurs in northern Luzon and the Bicol Region. Although northeastern Mindanao is an area with a low risk to typhoons, the heavy concentration of mining projects there generates concern when typhoons do impact that portion of the country.

The main risk posed by a typhoon to a mine concerns the stability of its tailings dam because the heavy rains associated with a typhoon can cause tailings impoundments to fail either though excess water pressure or by overtopping (Rico et al. 2008a, 2008b; World Resources Institute 2003). Worldwide, the largest cause of tailings dam failures are heavy rainfall events (Rico et al. 2008b). Many people in the Philippines familiar with either mining or natural hazards articulated their concerns about the dangers posed to mining by typhoons. Desiderio Cabanlit (the officer in charge of the Davao City Office of the Philippine Institute of Volcanology and Seismology PHIVOLCS), Rovik Obanil (the communications and networking officer of the Legal Rights and Natural Resources Center, an NGO engaging in advocacy on behalf of indigenous peoples affected by mining) and Ricardio Saturay (the program coordinator at the Center for Environmental Concerns (CEC), an environmental NGO), all regard meteorological hazards such as typhoons as the biggest risk to mines in the Philippines (Cabanlit 2007; Obanil 2009; Saturay 2009). According to Engineer Virgilio Perdigon, secretary general of Aquinas University and spokesperson for the Save Rapu-Rapu Alliance (an NGO opposed to mining on Rapu-Rapu Island) the numerous powerful typhoons containing tremendous amounts of moisture pose a serious threat to mines (Perdigon 2009). On 26 September 2009, Typhoon Ondoy deposited over 340 millimeters of rain on Metro Manila, and rainfall events of this magnitude can cause tailings dams to fail (Perdigon 2009). Jesus Garganera, the national coordinator of the Alyansa Tigil Mina (Alliance to Stop Mining, or ATM), views the most serious risk to mining in terms of frequency of occurrence (especially in Northern Luzon and in Bicol) as typhoons, (Garganera 2009).

There have indeed been many examples of typhoons, and heavy rainfall events, adversely impacting mines in the Philippines. In 1911, a strong typhoon struck the Baguio mining district causing landslides and high water; mining equipment was badly damaged and Benguet Consolidated Mining's mill and

Figure 4.1. The overlap of mining projects at risk from typhoons

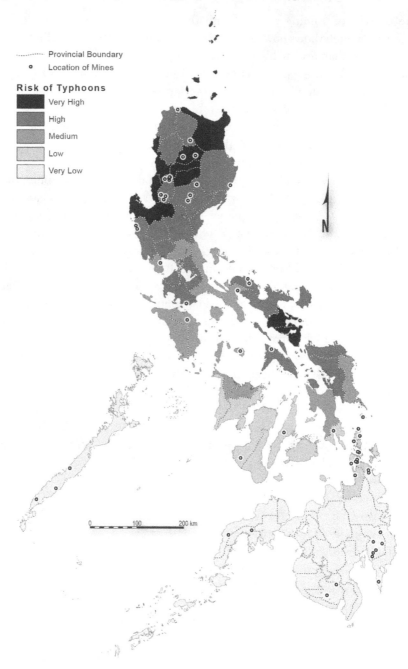

Source: Based on data from Manila Observatory (2010) and the Mines and Geosciences Bureau (2006, 2007a, 2007b, 2007c).

cyanide plant were destroyed, forcing the company to declare bankruptcy (Lopez 1992).

In 1964, flooding from typhoons affected the Toledo Mine on the island of Cebu; the flooding forced Atlas Consolidated Mining (the mine's operators) to halt production (United States Bureau of Mines 1965). In 1967 extremely heavy rains during a typhoon caused a mudflow that buried 43 workers alive at the Santo Thomas Mine in Benguet; of the 43 workers only 23 survived (Lopez 1992). Palawan Consolidated Mining's chromite mine on the island of Palawan was damaged by a typhoon in 1969 (United States Bureau of Mines 1970).

In 1976 Typhoon Didang caused serious damage to the chromite mine operated by Atlas Consolidated Mining at Coto, Zambales, on the island of Luzon and disrupted the road connecting it to the seaport (United States Bureau of Mines 1977).

From December of 1980 until February of 1981, the copper mine of Sabena Mining Corporation in New Bataan, Davao del Norte had to suspend operations due to flooding from heavy rains (Lopez 1992). A flash flood at the Amacan Copper Project in Masara, Davao del Norte (in 1981) resulted in the death of five workers (United States Bureau of Mines 1981). In 1984, production at the Nonoc Mining and Industrial Corporation's Surigao Nickel Complex was shut down due to extensive flooding caused by typhoons (United States Bureau of Mines 1984).

The operations of both the Benguet Corporation and the Philex Mining Corporation were severely disrupted by typhoons in 1990; the mines themselves were flooded, processing facilities were damaged and roads were blocked (United States Bureau of Mines 1990). That same year also saw the Siana Mine in Surigao del Norte experience a failure of its pit walls and flooding after exceptionally heavy rains associated with a typhoon (United States Geological Survey 2002). One year later the Benguet Corporation's Dizon Copper-Gold operation at San Marcelino, on the island of Luzon, was cut off due to flooding from torrential rains (United States Bureau of Mines 1991). In 1993 the Carmen Mine, operated by Atlas Consolidated Mining on the island of Cebu, was flooded during a typhoon (United States Bureau of Mines 1993). During the same year, heavy rains caused the tailings dam to overtop at the Itogon Suyoc Mines Itogon Gold Project – in Benguet Province in the Cordillera of Luzon (Landingin and Aguilar 2008). On 1 November 1996, heavy rains from a typhoon caused mine tailings from Maricalum Mining Corporation's tailings dams in Sipalay, Negros Occidental, to spill onto 500 hectares of rice fields after two of the company's tailings dams overflowed (Lansang 2011). This tailings spill affected 500 families and it was estimated that it would take between 10 to 12 years for their farmlands to recover

(Tujan and Guzman 2002). In 1997, the Dizon mine was forced to close due to a mudslide and flooding caused by a typhoon (United States Geological Survey 1997). One year later the Masinloc Mine operated by the Benguet Corporation on the island of Luzon experienced flooding caused by heavy rains; at the same time the Sibutad Mine operated by Philex Mining in Sibutad, Zamboanga del Norte experienced an overflowing of siltation resulting in an extensive fish kill (Philippine Indigenous Peoples Links 2007; United States Geological Survey 2000). In 1999, Manila Mining Corporation's Placer Mine, in Placer, Surigao del Norte, released 700,000 cubic meters of tailings during a heavy rainfall event (Landingin and Aguilar 2008).

The Dizon mine experienced additional problems in 2002 when heavy rains caused the erosion of a tailings dam and a spillage of tailings into a lake (Philippine Indigenous Peoples Links 2007). On 31 October 2005, tailings containing cyanide were spilled into Albay Gulf during a heavy rainfall event at the Rapu-Rapu Polymetallic Project; this followed another tailings spill occurring 20 days earlier as a result of a malfunctioning pump in the same tailings dam (Oxfam Australia 2008). September of the following year saw the power lines to the Rapu-Rapu Polymetallic Project disrupted by Typhoon Milenyo (*Mining Journal* 2006). Then, during November 2006, 11 people died when they were buried alive by a landslide during Typhoon Reming when it hit a mining affected community on Rapu-Rapu Island (Oxfam Australia 2008). On 11 July 2007, heavy rains caused the lower tailings storage facility to fill at the Rapu-Rapu Polymetallic Project, thus necessitating the construction of an emergency drainage canal (Philippine Indigenous Peoples Links 2007). Lastly, during February of 2011, the main settling pond[2] of the San Roque Metals La Fraternidad Project, in Tubay, Agusan del Norte, overflowed after several days of heavy rainfall spilling an estimated 20,000 liters per second (Arguillas 2001a). This led to a flooding of lowland areas, which was described as "unprecedented in the history of the municipality," with floodwaters in some areas reaching almost two meters in height (Arguillas 2001a). Two residents of Tubay – a 7-year-old boy and a 70-year-old man – drowned during the flooding. During the heavy rains both the municipal disaster risk reduction and management council and the local chief of police attempted to examine the state of the settling pond but were denied entry by the mining company.

Typhoons and climate change: An increasing risk

Not only have typhoons adversely impacted mines, it is a distinct possibility that the severity of typhoons will worsen as the atmosphere warms due to climate

---

2 Settling ponds are supposed to prevent silt from escaping into the natural waterways.

change occasioned by anthropogenic greenhouse gas emissions (Blanco 2006; Cutter et al. 2007; Wisner et al. 2004). There is no doubt remaining as to the validity of scientific studies indicating that the world's climate is warming. "An increasing amount of scientific information has been accumulated that supports the contention that contemporary global warming is human induced due to greenhouse gas emissions rather than being part of a natural climate variability cycle" (Yumul et al. 2011, 371). There is "ample evidence that continued increases in global greenhouse gas concentrations are likely to result in increases in the occurrence and intensity of extreme [high temperature] events" (Diffenbaugh and Scherer 2011, 2). Furthermore, the emissions of greenhouse gases into the environment over the last two hundred years have been so extensive that the world's climate will continue to warm even if greenhouse gas emissions are dramatically reduced. According to Yumul et al. (2011, 1): "even if the world makes a significant reduction in greenhouse gas emissions, the lag in the climate system means that the world is faced with decades of climate change due to the emissions already put into the atmosphere."

As the world's climate changes, the variability of weather will become more prevalent and there will be more extreme events such as tropical cyclones due to increased energy within the climate system (Cutter et al. 2007; Helmer and Hilhorst 2006; O'Brien et al. 2006; Thomalla et al. 2006). Van Aalst (2006, 12) wrote, "Global warming is already changing environmental conditions in the areas where tropical storms occur, providing more energy to fuel the storms, which can make them more intense." The most immediate increase in extreme heat is predicted to occur in the tropics where up to 70 percent of all seasons during the time period 2010–39 will have seasonal maximums in excess of the maximum seasonal temperatures occurring during the late twentieth century (Diffenbaugh and Scherer 2011). Research conducted by Kerry Emanuel, professor of atmospheric science at the Massachusetts Institute of Technology, indicates that worldwide, the destructiveness of tropical cyclones has increased over the past 30 years due to an increase in their average intensity and lifetime (Emanuel 2005, 2007). In addition to creating warmer sea temperatures (the principal cause of typhoons), climate change (by melting glaciers and the polar ice caps) will cause a rise in the world's sea level by between 15 to 59 centimeters over the next 100 years (Department of Environment and Natural Resources Climate Change Office 2010; Emanuel 2007). With higher sea levels, the storm surge prevalent during a typhoon will be even more intense and dangerous. This is an acute concern in the Philippines because, according to the Department of Environment and Natural Resources Climate Change Office (2010, 12), "based on elevation and population exposure, the Philippines [has come] out as one of the ten countries in the world which are highly vulnerable to sea level rise."

There are many in the Philippines who echo these concerns about climate change, amplifying the archipelago's vulnerability to typhoons. Amalie Obusan, a climate and energy campaigner for Greenpeace Southeast Asia, has no doubt that the two typhoons (Ondoy and Pepeng) that ravaged the Philippines at the end of September 2009 were the result of climate change due to the amounts of rain that fell relative to historical data (Obusan 2009). On 26 September 2009, Metro Manila was hit by Typhoon Ondoy and over 340 millimeters of rain fell in six hours; the previous 24-hour rainfall record was 340 millimeters, set 42 years earlier in 1967 (Obusan 2009).[3] Obusan recalled how the Philippine Red Cross asked Greenpeace Southeast Asia to use its inflatable boats (usually used for campaign purposes) to rescue people from the flooded streets of Marikina City in Metro Manila; "these typhoons were unlike anything ever seen in Philippine history" (Obusan 2009, interview). Lyra Magalang, the Oxfam Great Britain program officer for disaster risk reduction in the Philippines, is convinced that climate change is increasing the severity and frequency of typhoons and Magalang points to the fact that during September 2009 there were five typhoons during just one month (Magalang 2009). Carlos Conde, a journalist who has written about climate change for the *International Herald Tribune*, clearly attributes these typhoons to climate change as the volume of rain was simply incredible (Conde 2009). He finds it difficult not to attribute such powerful typhoons to climate change and views the combination of the archipelago's vulnerability to typhoons with climate change as a situation where "there are bound to be problems" (Conde 2009, interview).

An augmentation of typhoon severity and frequency as a result of climate change poses a serious problem for mining. As indicated in Chapter Three, the geochemical process of acid generation is something operating on a geologic time scale so any tailings dam designed to contain acidic tailings will require perpetual attention. If the government continues its aggressive promotion of mining, there will be numerous tailings dams distributed across the landscape of the archipelago. These tailings dams will also be of varying ages with those associated with operating mines being quite new and those associated with mines that have long since ceased production being essentially abandoned. Combine these tailings dams with the increased risk of accidents emerging from stronger and more frequent typhoons and the situation begins to resemble what Engineer Virgilio Perdigon likened to a "Sword of Damocles hanging over the Philippines" (Perdigon 2009, interview). In the opinion of Ricardio Saturay, with climate change we can expect more powerful and more frequent

---

3 With 340 millimeters of rain falling in only one-fourth of the time that it previously took 340 millimeters of rain to fall, the rainfall during this storm was four times higher than the previous 24-hour rainfall record.

typhoons; tailings impoundments need to be designed to take into account the greater intensity and frequency of typhoons (Saturay 2009). The destruction of mining infrastructure designed according to climatic data obtained prior to the current projections of climate change could cause massive amounts of pollution and damage to low lying areas with significant contamination of aquatic resources (Department of Environment and Natural Resources Climate Change Office 2010).

Earthquakes: Extreme Seismic Risk

Earthquakes: A release of geologic energy

An earthquake can be defined as "a trembling or shaking of the ground that is caused by the sudden release of energy stored in the rocks beneath the earth's surface" (Plummer and McGeary 1982, 304). The source of seismic activity in the Philippines is, unfortunately, the same source of the archipelago's rich mineralization: the juxtaposition of the Philippine Plate between the eastward moving Eurasian Plate, the northward moving Indo-Australian Plate and the westward moving Pacific Plate (Figure 1.7). Wedged between these much larger tectonic plates, the small Philippine Plate is an area of substantial seismic activity (Bankoff 2003a). As these plates are subducted beneath the Philippine Plate, friction is generated and the imperceptibly slow rate of subduction (occurring over geologic time) comes to a virtual halt (Punongbayan 1994). However, from time to time frictional resistance may be abruptly overcome and the downward moving plate lurches forward in a sudden motion transcending geologic time, producing shallow-seated earthquakes (Punongbayan 1994). The magnitude of earthquakes is displayed on the Richter scale which is a logarithmic scale of earthquake magnitude ranging between 0 and (theoretically) 10; this means that the difference between any two consecutive whole numbers on the scale means that there has been a ten-fold increase in the amplitude of the earth's vibrations (Plummer and McGeary 1982).[4]

According to Desiderio Cabanlit from PHIVOLCS, earthquakes can have three main effects: ground shaking, the disruptive up and down and sideways shaking motions; ground rupture, the creation of new (or the renewed

---

4 It has been estimated that a ten-fold increase in the size of the earth's vibrations is accompanied by an increase of about 31.5 times as much energy; this means that an earthquake of magnitude 5 will release 31.5 times as much energy than an earthquake of magnitude 4 while, an earthquake of magnitude 6 releases almost one thousand times as much energy (31.5 × 31.5 = 992.25) than an earthquake of magnitude 4 (Plummer and McGeary 1982).

movements of old) fractures, often with the two blocks on both sides moving in opposite directions; and liquefaction, a process transforming the behavior of a body of sediment from that of a solid into that of a liquid (Cabanlit 2007). Earthquakes may also occur in "swarms," where a series of earthquakes occur in a cluster in one area in a short period of time (MindaNews 2008). From 13 March to 14 March 2008, for example, 48 earthquakes were recorded in the province of North Cotabato on the island of Mindanao; one of these earthquakes registered at magnitude 1 on the Richter scale while two of these earthquakes registered at magnitude 5 (MindaNews 2008).

Vulnerability of the Philippines to earthquakes

The Philippines are highly vulnerable to earthquakes (Bankoff 1999; Luna 2001). The Pacific Ring of Fire generates approximately 80 percent of the world's earthquakes and the archipelago is located precisely in this area (Delica 1993). "Between 1950 and 1975," wrote Bankoff (2003a, 32), "the islands experienced 2,126 recorded earthquakes." Some of these earthquakes have been extremely powerful and extremely destructive. The 17 August 1976 Moro Gulf earthquake registered at magnitude 8 on the Richter scale; its damaging effects were felt as wide as Zamboanga City (145 kilometers to the west), Davao City (250 kilometers to the east) and Pagadian City (70 kilometers to the north) (Haas 1978). The 21 July 1990 Luzon earthquake registered at magnitude 7.8 on the Richter scale; it resulted in the deaths of 1,666 people and caused damages of USD 305 million (Luna 2001).

Earthquakes risk is a spatially variable phenomenon and some parts of the archipelago are more prone to earthquakes than others. The portions of the Philippines where earthquakes are most frequent are northern Luzon, around Manila, the southern tip of the Bicol Region and a large area of territory on Mindanao (shaped like a reversed "L") ranging from Surigao del Norte in the northeast to South Cotabato in the southwest. Portions of the archipelago where earthquakes are still frequent include Mindoro, the northern portion of Panay, Bohol and the Zamboanga Peninsula of Mindanao. Earthquakes are considered rare or infrequent on Negros, Cebu and Palawan. Of all the major regions, "Mindanao has been the most seismically active among the major island groups with 49.5 percent of the total number of medium-sized events occurring in the archipelago" (Mangao et al. 1994, 41).

Mining and earthquakes

The overlap of mining projects with the risk of earthquakes is presented in Figure 4.2 and this demonstrates that 42 percent of all mining projects in the Philippines are in areas that are at high risk of earthquakes, most notably

Figure 4.2. The overlap of mining projects at risk from earthquakes

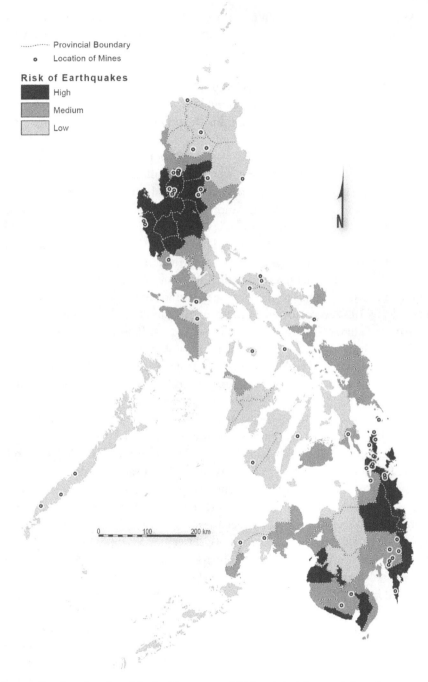

Source: Based on data from Manila Observatory (2010) and the Mines and Geosciences Bureau (2006, 2007a, 2007b, 2007c).

the Cordillera of northern Luzon and along the eastern side of the island of Mindanao.

Where the location of mining projects in areas prone to earthquake risk becomes particularly serious is along the eastern side of Mindanao, what Mitchell and Leach (1991, 194) referred to as the "Eastern Mindanao Gold Province." According to Romie Valerio, supervising geologist of the Davao MGB Office, Mindanao has the highest mineral potential of all of the Philippines and the best mineral potential on Mindanao is on the eastern side of Mindanao (Valerio 2005). While this portion of Mindanao may have the highest mineral potential in the country, it is also – according to Desiderio Cabanlit – "the most seismically active portion of the country" (2007, interview). This means that there will be a substantial number of mining projects located in an area with a high potential for seismic risk. "Because earthquakes can disrupt buildings and other infrastructure, they are a major safety concern for mine development" (World Resources Institute 2003, 30).

As with a typhoon, the main risk posed by to a mine by an earthquake concerns the stability of its tailings dam because the violent shaking associated with an earthquake can cause tailings impoundments to fail. Worldwide, the second most common cause of tailings dam failures are seismic events (Rico et al. 2008b). For example, in 1928 an earthquake measuring magnitude 8.3 on the Richter scale caused the failure of the waste impoundment at the Barahona Copper Mine in Chile and almost 3 million cubic meters of waste was released resulting in the deaths of 54 people (World Resources Institute 2003). As Dobry and Alvarez note, "tailings dams are dangerous structures and are of great seismic instability" (1967, 259). What makes tailings dams vulnerable to failure during an earthquake are the fine-grained nature of the tailings. Being fine wet sand, tailings are likely to liquefy when subjected to an earthquake and their behavior will abruptly change from being that of a solid into that of a liquid (Dobry and Alvarez 1967; Rimando 1994). In the opinion of Desiderio Cabanlit, it is not safe to conduct mining in areas prone to earthquakes although engineering may ameliorate the risks posed by them (Cabanlit 2007). To Jesus Garganera from the ATM, the most serious risk to mining in terms of frequency of occurrence may come from typhoons but the most serious risk to mining in terms of event magnitude comes from earthquakes (Garganera 2009). Engineer Virgilio Perdigon pointed out how Rapu-Rapu Island has two fault lines running through it and in both 2008 and 2009 local earthquakes occurred in the Bicol Region registering from between 3 to 5 on the Richter scale (Perdigon 2009). To Perdigon, there is a serious risk that an earthquake may cause a tailings dam failure at the Rapu-Rapu Polymetallic Project (Perdigon 2009). Mayor Jerry de la Cerna, the mayor of the municipality of Governor Generoso in the province of Davao Oriental on the island of Mindanao, indicated how Governor Generoso experiences up to ten earthquakes a year and, accordingly,

Mayor De la Cerna is concerned about one of these earthquakes impacting a mine and damaging its tailings dam (De la Cerna 2005).

There are examples of earthquakes impacting mines. In 1990, an earthquake damaged the Acupan Mine operated by the Benguet Corporation on the island of Luzon; electrical power lines were brought down and a landslide blocked access roads leading to the mine (United States Bureau of Mines 1990). In 1992, the Santo Thomas II Copper Mine, also on the island of Luzon, experienced a tailings pond collapse due to a weakening of its tailings dam during the 1990 Luzon earthquake (Landingin and Aguilar 2008). In 1995, Manila Mining Corporation's Placer Mine in Placer, Surigao del Norte, suffered a collapsed tailings pond due to tectonic movement (Landingin and Aguilar 2008). On 17 March 1996, a magnitude 3.2 earthquake occurred 5 kilometers east of the Marcopper Mine on the island of Marinduque (Futures Group International 2004). While this minor earthquake occurred long after the 25 August 1995 onset of seepage emanating from the Tapian pit, "it is possible that this seismic event could have aggravated the situation," contributing to the worst mining related environmental disruption in the Philippines seven days later on 24 March 1996 (Futures Group International 2004, 3–10). Clearly, the risk posed by earthquakes to mines is a real and extant risk and is more than a hypothetical possibility.

Tsunamis: Seismically Induced Floods

Tsunamis: Seismic sea waves

Tsunamis are seismic sea waves caused by earthquakes that disturb the sea floor (Plummer and McGeary 1982). When a large section of sea floor is suddenly raised or lowered by an earthquake occurring beneath the sea floor, the water over the moving area will be lifted (or dropped) during that earthquake. Then when the water returns to sea level it will generate a long low wave that spreads quickly over the ocean. Tsunamis may have a length of 150 kilometers and travel at speeds up to 750 kilometers an hour. While the height of a tsunami wave in the open ocean may only be 1 to 3 meters, its height will rise as it approaches shore reaching up between 15 to 30 meters, and when a tsunami hits shore it will capable of causing substantial damage (Plummer and McGeary 1982).

Tsunamis in the Philippines

Being located amid the "Pacific Ring of Fire," the islands are vulnerable to tsunamis originating as far away as Japan or even Chile. Tsunamis have periodically ravaged Philippine coasts; since 1603, there have been at least twenty-seven tsunamis with waves in excess of 25 meters that have impacted the coastline of the archipelago (Bankoff 2003a). The 17 August 1976 tsunami which impacted coastal areas of the Moro Gulf was a catastrophic tsunami with waves up to 9 meters high; this caused between 3,000 and 6,500 deaths, injured 8,000 people and left 12,000 families homeless (Bankoff 2003a).

Mining and tsunamis

The overlap of mining projects with the risk of tsunamis is depicted in Figure 4.3 and this indicates that only 23 percent of all mining projects in the Philippines are in areas that have no risk of being impacted by a tsunami. While there are currently no examples of tsunamis adversely impacting mines in the Philippines, there were documented instances wherein mine pits were flooded in Thailand during the 26 December 2004 tsunami that impacted that nation (Umitsu et al. 2007).

Volcanoes: An Additional Source of Seismic Activity

Volcanoes: Extrusive igneous phenomenon

Volcanoes are hills or mountains constructed by the extrusion of lava or rock fragments from a vent (Plummer and McGeary 1982). While some volcanoes, such as the Hawaiian Islands, are formed as a tectonic plate moves over a hot spot (or mantle plume) in the earth's outer core, most volcanoes occur at plate boundaries (Punongbayan 1994). As one plate is subducted beneath another plate, the rocks of the plate being subducted melt and are turned into magma. This rising magma (as explained in Chapter One) is what creates ore deposits and when magma accumulates within a magma chamber inside the lithosphere, volcanoes are created. When the pressure within that magma chamber reaches such a point, it suddenly explodes and eruption occurs.

Volcanoes in the Philippines

In the Philippines there are 220 volcanoes and 22 of these are considered active (Delica 1993; Punongbayan 1994). It bears stressing, however, that the definition of "active" and "inactive" volcanoes is an imprecise concept. Mount Pinatubo was considered inactive for centuries but after the 21 July 1990 Luzon earthquake, it became active again and violently erupted in June 1991 (Bankoff 2003a). Volcanoes are what Bankoff

Figure 4.3. The overlap of mining projects at risk from tsunamis

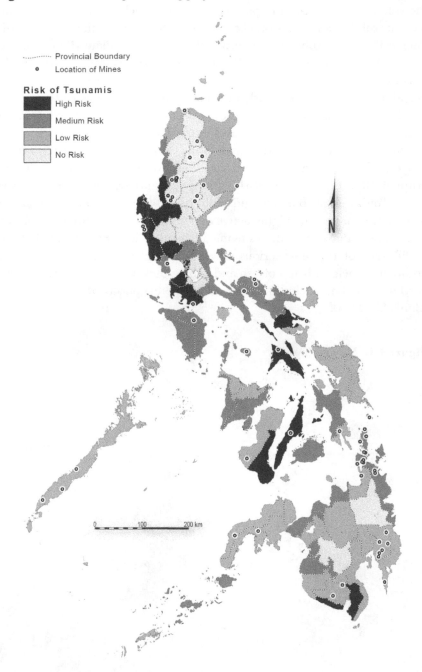

Source: Based on data from Manila Observatory (2010) and the Mines and Geosciences Bureau (2006, 2007a, 2007b, 2007c).

(2003a, 38) refers to as "a feature of the Philippine landscape." Most of the volcanoes in the country parallel deep sea trenches associated with the tectonic subduction zones bracketing the archipelago as the Eurasian and Pacific Plates are subducted beneath the Philippine Plate (Punongbayan 1994). The volcanoes in the Bicol Region, for example, parallel the northwest trending Philippine Trench (Punongbayan 1994). A volcanic eruption can be a spectacularly destructive event; the June 1991 eruption of Mount Pinatubo was one of the most violent and destructive volcanic events of the twentieth century and more than 5 billion cubic meters of ash and debris were ejected (Fujikura and Nakayama 2001, 192). Mount Mayon (Figure 4.4) is the most active volcano in the Philippines, having erupted 45 times between 1616 and 1985 and will erupt once every 8 years on average (Bankoff and Hilhorst 2009, 692). During late 2009, Mount Mayon became highly active and it appeared that an eruption was so imminent that communities living adjacent to it were evacuated (Nasol 2009). One of the most serious aspects of a volcanic eruption is that in addition to their emission of ash and debris, they are often associated with earthquakes. This includes almost all of the eruptions of Mount Mayon (Bankoff 2003a).

Figure 4.4. Mount Mayon

Volcanoes and mining

In Figure 4.5, the overlap of mining projects with the risk of volcanic eruptions is presented and this illustrates how almost one-third of those in the Philippines are located in areas where there is some degree of risk to volcanic eruptions. In the case of the Rapu-Rapu Polymetallic Project, there is a very high risk of a volcanic eruption as Mount Mayon is only 49 kilometers away (Figure 4.6).

In the event of a large-scale volcanic eruption such as the eruption of Mount Pinatubo in 1991, probably the last thing anyone would think about would be its impact on a mine as the overall consequences of such an eruption would be so catastrophic that the damages inflicted upon a mine would seem almost inconsequential. Nevertheless, volcanoes do pose a substantial hazard to mining projects. Ricardio Saturay, a geologist, indicated that the ash from a volcano could cause critical mining project machinery to become clogged and be rendered inoperative and that volcanic eruptions can generate volcanic tremors capable of damaging a tailings dam (Saturay 2009). Engineer Virgilio Perdigon expressed his concern that the two fault lines running through Rapu-Rapu Island could interact with Mount Mayon and a volcanic tremor could occur, adversely affecting the Rapu-Rapu Polymetallic Project (Perdigon 2009). In 1991, the accumulation of volcanic ash following the eruption of Mount Pinatubo caused mining to be suspended at the Dizon Copper-Gold mine operated by the Benguet Corporation on the island of Luzon (United States Bureau of Mines 1991). Although not as widespread a hazard as typhoons and earthquakes, volcanoes also pose a risk to mining operations.

El Niño–Induced Drought: An Abrupt Decrease in Rainfall

The El Niño Southern Oscillation: An alteration of global weather patterns

The El Niño Southern Oscillation (ENSO) is a naturally occurring phenomenon wherein global weather conditions suddenly, and for several months, depart from their normal patterns and abnormal weather appears over large portions of the world. The El Niño phenomenon has its origins in the waters off of Peru in the southeastern Pacific Ocean where the Peru Current[5] flows from south to north bringing cold Antarctic water up into the tropics. Normally this cold current will cause rich nutrients to up-well from deep within the ocean, thus creating a thriving fishing industry in communities on the west coast of South America such as those in Chimbote, Peru. This cold current is

---

5 The Peru Current is also called the Von Humboldt Current after Alexander von Humboldt, the famous nineteenth-century Prussian naturalist.

Figure 4.5. The overlap of mining projects at risk from volcanic eruptions

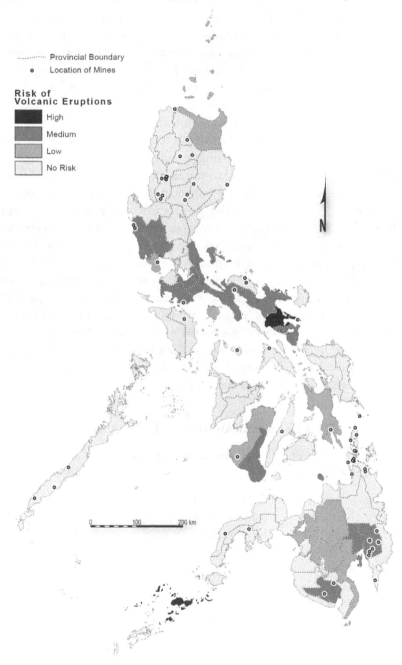

Source: Based on data from Manila Observatory (2010) and the Mines and Geosciences Bureau (2006, 2007a, 2007b, 2007c).

Figure 4.6. The proximity of Mount Mayon to the Rapu-Rapu Polymetallic Project

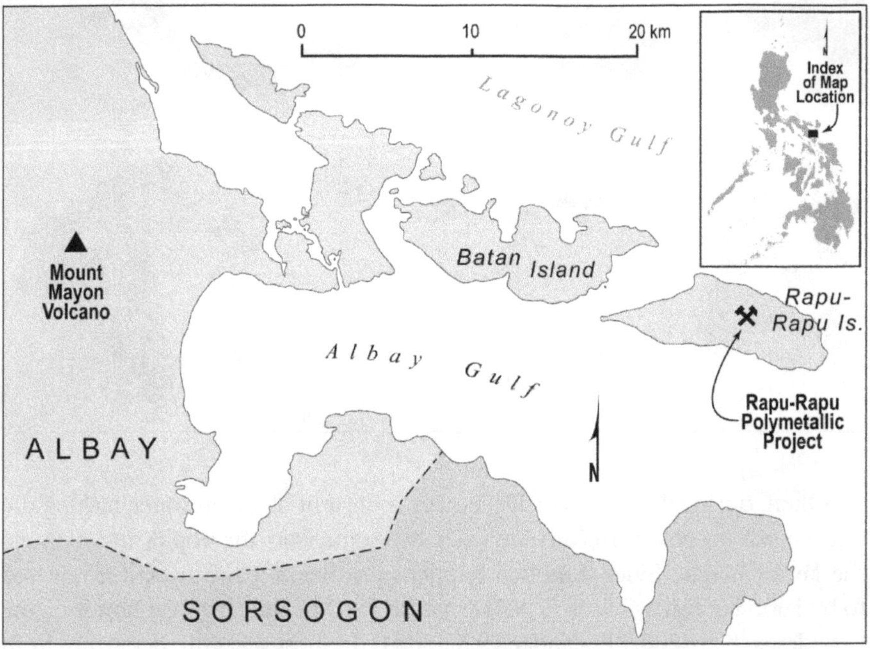

associated with high air pressure which, by suppressing precipitation, creates the desert-like climatic conditions found along the Peruvian coast. When the Peru Current interacts with the prevailing trade winds blowing from east to west in equatorial latitudes, there will be low air pressure prevailing in the western Pacific Ocean. Since this low air pressure does not suppress precipitation, places in the western Pacific – such as the Philippines – often experience high levels of precipitation. This pattern of high air pressure (with low levels of precipitation) in the eastern Pacific and low air pressure (with high levels of precipitation) in the western Pacific is one of the great climatic constants in the tropical Pacific (Table 4.1 and Figure 4.7).

Table 4.1. Normal climatic conditions in the tropical Pacific

Western Pacific Ocean	Eastern Pacific Ocean
Low air pressure	High air pressure
High levels of precipitation	Low levels of precipitation
High sea levels	Low sea levels

Source: Bankoff (2003a), Katz (2002) and Philander (1990, 2004).

Figure 4.7. Normal climatic conditions in the tropical Pacific

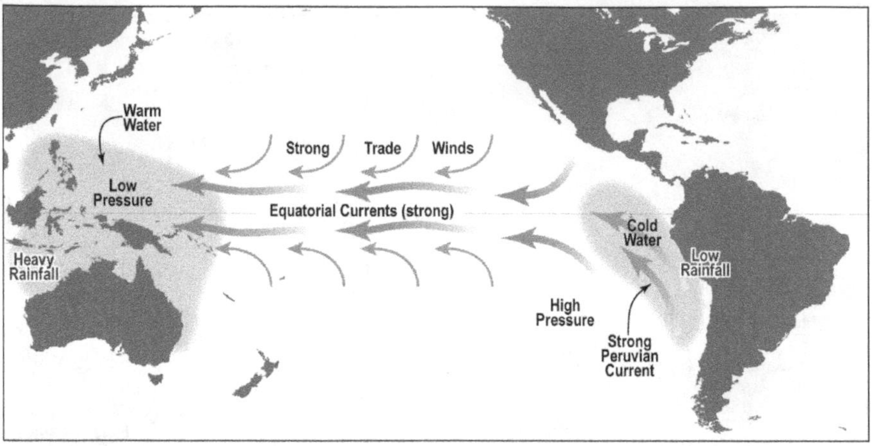

Often, during the month of December, a current of warm water, lacking the nutrients of the cold Peru Current, will move south from the tropics and suppress the Peru Current. Since this often happens just before Christmas, it is referred to by Peruvian fisher-folk as *El Niño* (Spanish for "the little boy") in honor of the baby Jesus (Katz 2002; Philander 1990, 2004). In most years this departure from the normal lasts around thirty days until the cold Peru Current can reassert itself (Katz 2002; Philander 1990, 2004). However, once every three to seven years, this phenomenon becomes very strong and lasts for several months and in this case an El Niño event is said to be underway (Katz 2002; Philander 1990, 2004). During an El Niño, event conditions in the tropical Pacific reverse themselves and what is normal in the eastern Pacific becomes prevalent in the western Pacific and what is normal in the western Pacific becomes prevalent in the eastern Pacific; hence the use of the term "El Niño Southern Oscillation" (ENSO) to describe an El Niño event (Katz 2002). During an El Niño, atmospheric pressures will shift and the westward blowing trade winds will slacken and even reverse themselves (Table 4.2 and Figure 4.8). Sea levels in the western Pacific will begin to fall and warm water will slip back towards South America disrupting the cold Peru

Table 4.2. Climatic conditions in the tropical Pacific during an El Niño

Western Pacific Ocean	Eastern Pacific Ocean
High air pressure	Low air pressure
Low levels of precipitation	High levels of precipitation
Low sea levels	High sea levels

Source: Bankoff (2003a), Katz (2002) and Philander (1990, 2004).

Figure 4.8. Climatic conditions in the tropical Pacific during an El Niño

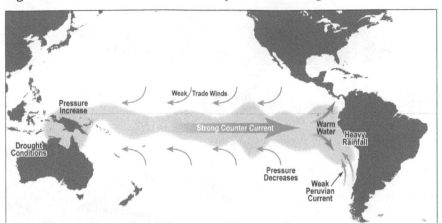

Current (Bankoff 2003a; Katz 2002; Philander 1990, 2004). As this normally cold current becomes disrupted, the eastern Pacific Ocean warms and air pressure becomes reduced with lower air pressure in the eastern Pacific extremely heavy rains will develop in normally arid Peru (Caviedes 1985; Van Aalst 2006; Warner and Ore 2006; Yumul et al. 2011). As extremely heavy rains develop in normally arid Peru colder drier air that normally descends over the eastern Pacific shifts further east and descends over Southeast Asia, lowering rainfall levels and triggering drought (Dawe et al. 2009; Rodolfo and Siringan 2006; Yumul et al. 2011). The Southeast Asian drought associated with El Niño can be so severe that the 2003 United Nations Development Programme (UNDP) Human Development Report identified El Niño as one of the two main factors, along with the East Asian financial crisis, contributing to the resurgence of poverty at the beginning of the millennium (Bello et al. 2009).

The effect of El Niño in the Philippines

In the Philippines there has been severe drought conditions associated with El Niño events during 1982–83, 1990–92, 1997–98, 2005–06 and 2009–10 (Balane 2010; Bankoff 2003a; Estabillo 2006, 2007a, 2007b; Flores 2010; Torion 2010). The 2010 El Niño–induced drought (Figure 4.9) was a serious event especially on the island of Mindanao, leading to power failures (as hydroelectric reservoirs were unable to maintain large enough supplies of water to generate electricity) and between USD 216 million to USD 432 million worth of agricultural production was lost (Balane 2010; Flores 2010; Torion 2010).

In Malaybalay, Bukidnon, precipitation during the first four months of 2010 was only 48 percent of its normal levels and Figure 4.10 contrasts the actual

Figure 4.9. Rice paddies in dried riverbeds in North Cotabato during the 2010 El Niño

Photo credit: Keith Bacongco

Figure 4.10. Precipitation in Malaybalay, Bukidnon, January to April 2010

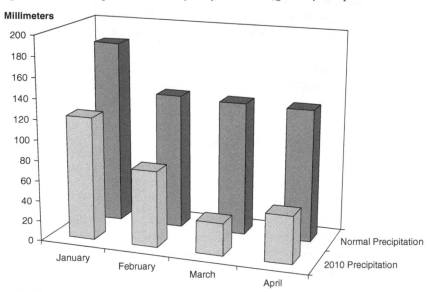

Source: Torion (2010).

precipitation during those months with what would normally be received in a non–El Niño year (Torion 2010).

The 1997–98 "mega-Niño" was an El Niño–induced drought of legendary proportions effecting five million people, especially on the island of Mindanao (Bankoff 2003a). The hardest hit parts of Mindanao were the provinces of North Cotabato, South Cotabato, Maguindano, Sarangani, Sultan Kudarat, Lanao del Norte and Lanao del Sur (Bankoff 2003a). Carlos Conde recalled how he had visited Mindanao during this drought and saw people who had died as a result of eating *kayos*, a type of root crop that must be soaked in running water before being cooked; water was unavailable for soaking the *kayos* and in the absence of any other type of food, these people ate the *kayos* out of desperation with tragic results (Conde 2009). After this drought, the government pursued a policy of extensive use of groundwater through the installation of shallow tube well pumps and by the end of 1998, had financed the construction of over 12,000 shallow tube wells (Bankoff 2003a; Rodolfo and Siringan 2006; Yumul et al. 2011).

El Niño–induced drought and mining

The overlap of mining projects with the risk of El Niño–induced drought is portrayed in Figure 4.11 that reveals that 28 percent of all mining projects in the Philippines are located in areas (Cebu, southern Leyte, and the northeastern portion of Mindanao) that are at high risk of drought during an El Niño event while 25 percent of all mining projects are located in areas (southern Palawan, and the remainder of Mindanao) that are at very high risk of being impacted by El Niño–induced drought.

Mining amid drought can generate substantial problems for both the operator of a mine as well as for communities near a mine. In the case of the former, a severe drought on the island of Cebu curtailed production at Atlas Consolidated Mining's Toledo Mine in 1970 (United States Bureau of Mines 1970). In the case of the latter, Emanuel Isip, the technical director of the Protected Areas and Wildlife Bureau (PAWB) of the DENR for southeastern Mindanao, related his experience as the Environmental Management Bureau (EMB) director for northeastern Mindanao when mine-pit dewatering at the Placer Mine of Manila Mining Corporation caused creeks to dry up (Isip 2005). Similarly, Mayor Jerry de la Cerna related how mine-pit dewatering due to historic chromite mining has resulted in reduced water flow in the municipality of Governor Generoso (De la Cerna 2005).

Concerns about the impact of mining upon groundwater become accentuated when one bears in mind the slow rate of aquifer recharge. Groundwater renews so slowly it should not be perceived as a renewable

Figure 4.11. Risk of decreased rainfall due to El Niño–induced drought

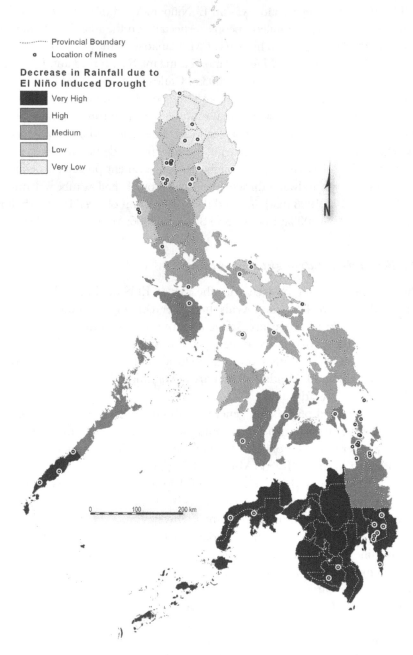

Source: Based on data from Manila Observatory (2010) and the Mines and Geosciences Bureau (2006, 2007a, 2007b, 2007c).

resource (something capable of being renewed in a timeframe meaningful to humans). Instead, groundwater renews so slowly it should be thought of as a nonrenewable resource (Bankoff 2003a). Extracting groundwater can never be more than a temporary measure to alleviate a crisis as its slow replenishment rate means resources are rapidly depleted. If there is an extensive El Niño–induced drought, the extensive use of shallow tube wells could lead to an exhaustion of groundwater. In the Philippines, the Water Code of 1976 (Presidential Decree 1067) requires all drilling permits to consider extracting water no faster than it can be naturally recharged. In practice, however, this requirement is routinely ignored and this code exempts all wells shallower than 10 meters (Rodolfo and Siringan 2006). Should there be an El Niño–induced drought leading to a reliance upon (and possible exhaustion of) groundwater resources, many areas across the archipelago will become extremely vulnerable to mining related hydrological disruption. Perhaps the most salient example of a mine with the potential to interfere adversely with water resources is the Tampakan Project, in the province of South Cotabato. This mine is projected to be a large mine that will be developed using open-pit methods (Estabillo 2009a). However, South Cotabato is an area that suffers from substantial decreases in rainfall during an El Niño (Bankoff 2003a). During a 2005 drought, approximately 2 million USD worth of crops were destroyed and it is feared that by 2070 the province's water resources will be exhausted (Estabillo 2006, 2007a, 2007b). Should this mine interfere with the groundwater of South Cotabato through mine-pit dewatering, (or through the extensive use of water for the purposes of ore treatment) it could exacerbate an already vulnerable water situation. By February 2010, the ATM, concerned about mining's impact on groundwater resources, called for the government to issue a moratorium on large-scale mining during El Niño events in order to mitigate the effects of drought (Alyansa Tigil Mina 2010). According to the ATM, provinces encountering drought during an El Niño will have better chances of coping with water stress if moratoriums on mining are imposed (Alyansa Tigil Mina 2010).

An interesting example of how El Niño (a global atmospheric phenomenon) adversely interacts with recent efforts by many developing countries to encourage mining (a global economic phenomenon) comes from northern Peru. In Peru, from 1999–2002, Manhattan Minerals, a Canadian junior mining company, attempted to develop an open-pit gold mine within the town of Tambogrande, Peru (Muradian et al. 2003). Northern Peru is a place where, during an El Niño event, there will be extremely heavy rainfall events with associated flooding (Caviedes 1985; Van Aalst 2006; Warner and Ore 2006; Yumul et al. 2011). Just as the Alyansa Tigil Mina in the Philippines currently opposes mining because an El Niño–induced drought can adversely impact mine-pit dewatering, members of the Frente de Defensor de Tambogrande

(Tambogrande Defense Front) in Tambogrande vigorously opposed the efforts of Manhattan Minerals to develop an open-pit gold mine in their town partly out of concern that the extremely heavy rains associated with an El Niño event would cause the mine, and its tailings dam, to flood and contaminate the rich agricultural lands surrounding Tambogrande (Maza 2002). The opposition of the Frente de Defensor de Tambogrande ultimately led to Manhattan Minerals abandoning their attempts to develop the mine (Renique 2006).

El Niño–induced drought and climate change

Anxiety over the effects of mining amid El Niño–induced drought only grow when one takes into account that like typhoons, the frequency and severity of El Niño events may also be increasing as a result of climate change (Department of Environment and Natural Resources Climate Change Office 2010; Emanuel 2007). According to Wisner et al. (2004, 83), "Considerable work over the past ten years on the El Niño Southern Oscillation suggest that these cycles of exceptionally wet and exceptionally dry weather, associated with periods of warming of surface water in the Pacific, may be increasing in frequency." With intensified El Niño–induced drought as a result of climate change, there will undoubtedly be increased threats to mine water supply security and increased competition between mine operators and nearby communities for water supply (Department of Environment and Natural Resources Climate Change Office 2010).

Synergistic Relations between Hazards

Synergistic effects between natural hazards

It should be stressed that the preceding hazards do not exist in isolation from each other and can often interact, making their combined effects substantially more serious. Heavy rains from a typhoon can increase the likelihood of an earthquake causing a tailings dam to fail because aseismic tailings dams require adequate drainage (Dobry and Alvarez 1967; Rico et al. 2008b). This is a serious concern because approximately 20 percent of the mining projects depicted in Figures 4.1 and 4.2 are located in areas that are at high risk for both typhoons and earthquakes. A tailings dam might be capable of withstanding an earthquake if filled with dry tailings but if filled with water, due to the heavy rains associated with a typhoon, its probability of failure will increase dramatically because the wet tailings will be more likely to liquefy (Dobry and Alvarez 1967; Rico et al. 2008b). This is what happened in 1995 at Manila Mining Corporation's Placer Mine (Landingin and Aguilar 2008). A tailings pond collapsed due to tectonic movement but the seismic event was preceded by a heavy rainfall event and the wet tailings liquefied very readily once the earthquake occurred. Groundwater withdrawal, necessitated by an El Niño–induced drought, can lead to land subsidence (Rodolfo and Siringan 2006). As water is progressively removed from an aquifer, the density of that aquifer diminishes and, over time, the level of the ground will be reduced. Combine land subsidence with higher sea levels brought on by the reduction of air pressure during a typhoon and the vulnerability of coastal areas to typhoons will increase even more with such areas becoming more susceptible to flooding brought on by the typhoon's storm surge. This generates a perverse situation where a lack of rain (a drought) leads to ameliorative measures (groundwater withdrawal) that increase vulnerability to the problems that occur when the rains return during typhoon season.

Natural hazards and anthropogenic environmental degradation

These natural hazards can also have a synergy with anthropogenic environmental degradation, the most notable example of such degradation being deforestation. The islands of the archipelago have experienced a substantial loss of their forest cover (Figure 4.12).

From 1575 until 1997, the Philippines experienced an annual deforestation rate of 0.39 percent (IBON 2006a). Deforestation accelerated substantially during the twentieth century. In 1900, old-growth rain forest covered approximately 70 percent of the national land area; by 1992, the percentage of land area covered by old-growth forest had been reduced to only 8 percent in scattered and fragmented patches distributed through the archipelago (Heaney and Regalado 1998). This deforestation can be attributed to several factors such as "the conversion of forests to agricultural land, commercial logging and the pressures of population growth" (Vitug 2000, 11).

Deforestation substantially augments other hazards with respect to landslides. According to Saldivar-Sali and Einstein (2007) there are four factors contributing to landslides: mountainous terrain, the heavy rainfall associated with typhoons, earthquakes and deforestation. Figure 4.13 depicts the overlap between mines and the percentage of each province consisting of land with a slope greater than 30 degrees, what Saldivar-Sali and Einstein (2007) regard as the gradient for classifying terrain as "steep."

Should there be a typhoon (with its associated heavy rains) or an earthquake (with its associated violent shaking of the ground) landslides will occur much more quickly than if the forest cover had been left on the hillsides (Heaney and Regalado 1998). As the Futures Group International (2004, 2–2) wrote, "Land surface instabilities caused by changing land use can accelerate slope failure during typhoons and earthquakes." Bear in mind that remoteness from

Figure 4.12. Percentage of land with forest cover

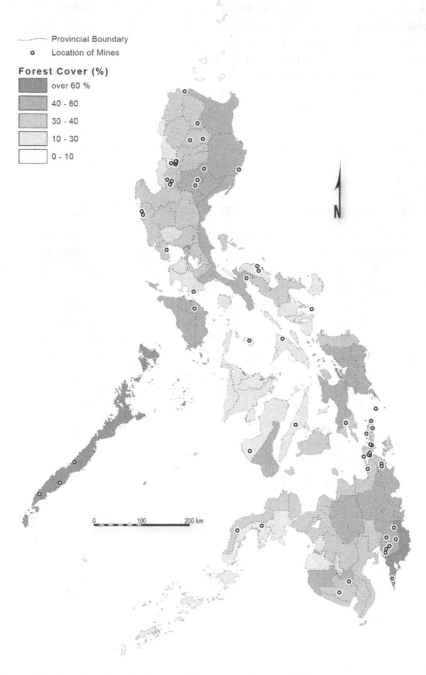

Source: Based on data from the Forestry Management Bureau (2010).

Figure 4.13. Topography and locations of major operating and proposed metallic mines

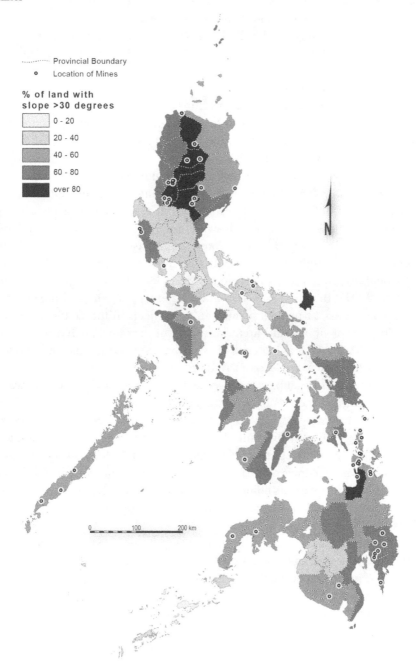

Source: Based on data from Manila Observatory (2010) and the Mines and Geosciences Bureau (2006, 2007a, 2007b, 2007c).

the ocean will fail to mitigate the effects of typhoons as the deforestation of coastal areas allows typhoons to penetrate further inland and inflict damage over wider areas than would formerly be the case (Myers 1988).

Landslides can pose major challenges for mining operations (Estabillo 2009b; Garcia 2006; MindaNews 2006; Morales 2008). In 1997, a landslide into a siltation pond at Philex Gold's Sibutad Mine in Sibutad, Zamboanga del Norte resulted in floods that destroyed nearby homes and rice fields (Philippine Indigenous Peoples Links 2007). Engineer Virgilio Perdigon is of the view that it is dangerous to locate mines in areas with steep slopes that have also been deforested because this creates a potential for landslides (Perdigon 2009). Mining in a deforested area – particularly if it involves the use of heavy machinery and blasting – could act as a triggering event, generating landslides (Morales 2008). According to Amalie Obusan, the September 2009 Cordilleran landslides which followed the passage of Typhoon Pepeng, occurred in areas where mining was being conducted and there are anecdotal reports that nitrates, from blasting explosives, could be smelled after the landslides (Obusan 2009). Indeed, the province of Benguet is particularly vulnerable to mining related landslides (Saldivar-Sali and Einstein 2007). Benguet, with less than 1 percent of all land area in the Philippines, hosts 14 percent of all mining projects in the archipelago, has over 80 percent of its land area consisting of terrain that has a slope in excess of 30 degrees, is at high risk of typhoons and earthquakes and has less than 40 percent of its forest cover remaining!

Deforestation also substantially contributes to drought because water that runs-off quickly does not enter the groundwater (Heaney and Regalado 1998). Where mines are located in areas of high deforestation, concerns about mining's impacts on groundwater resources also become amplified. In areas that are deforested and susceptible to El Niño, induced drought becomes an acute concern. The province of South Cotabato on the island of Mindanao is a case in point. This province has only 34 percent of its forest cover left and is also at very high risk of drought during El Niño (Bankoff 2003a; Estabillo 2007b).

Ultimately, it is quite conceivable that anthropogenic climate change will lead to more earthquakes, tsunamis and volcanism. It is well established that climate change will lead to a reduction of the thickness of the polar ice caps (Department of Environment and Natural Resources Climate Change Office 2010; Emanuel 2007). As the thickness of the polar ice caps is reduced, there will be a reduction of the weight being borne by the tectonic plates that rest upon the outer mantle. The corresponding isostatic adjustment of the outer mantle could generate more tectonic activity, thus more seismic and volcanic activity (Brandes et al. 2011; Galgana and Hamburger 2010; Grollimund and Zoback 2001; Poutanen and

Ivins 2010). Such effects may take several hundred years to manifest themselves but, as Chapter Three made clear, mining projects will require perpetual care and attention so even if enhanced seismic activity due to climate change does not manifest itself for hundreds of years, there will certainly be tailings ponds scattered across the islands of the Philippines that will become vulnerable to the enhanced seismicity brought about by anthropogenic climate change.

How These Hazards Can Create a Disaster

Disaster defined

The hazards discussed herein do not, in and of themselves, constitute disasters as no human beings are necessarily involved. Wisner et al. (2004, 50) define disaster as the situation when "a significant number of vulnerable people experience a hazard and suffer severe damage and/or disruption of their livelihood system in such a way that recovery is unlikely without external aid." To discuss disasters accurately, "the social production of vulnerability needs to be considered with at least the same degree of importance that is devoted to understanding and addressing natural hazards" (Wisner et al. 2004, 49). The essence of a disaster is the impact of a hazard upon vulnerable people; as Wisner et al. (2004, 87) wrote, "Disasters occur as the result of the impact of hazards on vulnerable people." A powerful typhoon (or tsunami) coming ashore on an uninhabited island would not constitute a disaster as no vulnerable people will be affected by it. Similarly, neither would a high magnitude earthquake occurring in the interior of Antarctica during the middle of the winter as no vulnerable people would be affected. To Wisner et al. (2004, 55) "it is the vulnerability of people that is crucial to understanding disasters." A relatively small hazard that impacts upon an extremely vulnerable population will produce a disaster; a substantial hazard impacting upon a population of low vulnerability will, nevertheless, produce a disaster if the magnitude of the hazard is sufficient. In either event, regardless of its magnitude, a hazard that has no impact upon any vulnerable people (the hypothetical example of an earthquake occurring in the middle of the Antarctic winter) will not constitute a disaster.[6]

---

6 The concept of a disaster may be explained by the pseudoequation: $D = H \times V$. Disaster (D) is the combination of a hazard (H) with a vulnerable population (V). In this pseudoequation, the mathematical operation of multiplication is used because the two terms combine with each other as opposed to simply summing. This is done because there must be both a hazard and a vulnerable population. Should either of these be missing there will be no disaster because the product of any number with zero will always be zero no matter how large that number may be (Wisner et al. 2004).

The concept of vulnerability

If the essential averment of defining a disaster is the impact of a hazard upon a vulnerable population one may reasonably ask, what constitutes vulnerability? To Cutter (1996, 529), vulnerability may be "broadly defined as the potential for loss." Wisner et al. (2004, 11) define vulnerability as "the characteristics of a person or group and their situation that influence their capacity to anticipate, cope with, resist and recover from the impact of a natural hazard." The Department of Environment and Natural Resources Climate Change Office (2010, 34) describes vulnerability as "the capacity to be wounded."

Poverty: An important determinant of vulnerability

One of the most important characteristics influencing the capacity of people to cope with a hazard is poverty. The poor and marginalized tend to have a low capacity to cope with a hazard and consequently, are highly vulnerable (Bankoff 2003a; Bankoff and Hilhorst 2009; Cutter et al. 2003; Delica 1993; Wisner et al. 2004). "Vulnerable populations are those at risk, not simply because they are exposed to hazard, but as a result of marginality that makes of their life a 'permanent emergency'" (Bankoff 2003a, 12). Poverty "is a primary contributor to social vulnerability as fewer individual and community resources for recovery are available, thereby making the community less resilient to the hazard's impacts" (Cutter et al. 2003, 251). People become vulnerable when they have "inadequate livelihoods which are not resilient in the face of shocks" (Wisner et al. 2004, 56).

The Philippines: Spaces of vulnerability

The high population density of the archipelago

As indicated in Chapter One, the islands of the Philippines are spaces of vulnerability. As Bankoff and Hilhorst (2009, 689) wrote: "Disasters are a fact of life in the Philippines. A socially and economically vulnerable population combines with one of the world's most hazardous landmasses to make disasters a frequent life experience." The first source of vulnerability in the Philippines is the high population density of the archipelago. In Figure 4.14, the population density of the archipelago is compared with the population densities of the top five destinations for mining investment in the 2010/2011 Fraser Institute Survey of Mining Companies: Chile, Quebec, Saskatchewan, Nevada and Greenland (Fraser Institute 2011; New Internationalist 2007; United States Census Bureau 2011).

Figure 4.14. Population densities: The Philippines and top five Fraser Institute jurisdictions

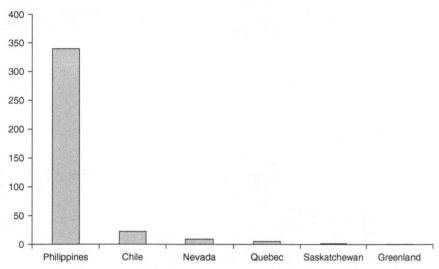

Source: United States Census Bureau (2011).

The population density of the Philippines is approximately forty-four times higher than that of the average population density of these jurisdictions; this means that there are substantially more people capable of being impacted by a mining related environmental disruption in the Philippines than there are in the top five destinations for mining investment.

Poverty in rural areas

Not only are there large numbers of people living in the Philippines, there are also high rates of poverty in rural areas where mining projects are located.[7] In Figure 4.15, mine locations are overlapped upon the rural poverty rates shown in Figure 1.3 and there are many provinces in the Philippines where there are high rates of rural poverty as well large numbers of mining projects.

Take into account the province of Surigao del Norte on the island of Mindanao. This province has a rural poverty rate of approximately 60 percent and it hosts almost 16 percent of all mining projects in the country even though it has less than 1 percent of the archipelago's land area. This overlap

---

7 Generally as a rule mining projects are located in rural areas, not in preexisting urban areas. Although there have been instances in other countries where mining projects have been slated for development within urban areas, (such as the unsuccessful efforts of Manhattan Minerals to develop their mine in the town of Tambogrande, Peru) the authors know of no such instances in the Philippines.

Figure 4.15. Rural poverty rates and location of mining projects

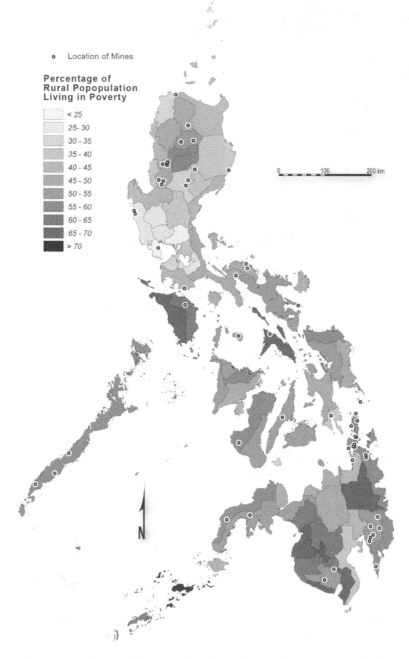

Source: Based on data from the Mines and Geosciences Bureau (2006, 2007a, 2007b, 2007c) and National Statistical Coordination Board (2006).

of rural poverty with mining means that there is a high likelihood of the rural poor being adversely affected in the event of a mining related environmental disruption. Couple the fact that Surigao del Norte is a high risk area for both earthquakes (Figure 4.2) and El Niño–induced drought (Figure 4.11) and the likelihood of the rural poor being adversely impacted by a tailings dam failure or water table drawdown seems event more apparent.

The reliance of the rural poor upon subsistence agriculture and aquaculture

As indicated in Chapter One, the vulnerability of the rural poor emanates from their heavy reliance upon subsistence agriculture and subsistence aquaculture. People who engage in such activities are highly vulnerable to any form of environmental disruption. As subsistence farmers and fisher-folk, poor people's livelihoods depend directly on their access to natural resources. Such dependence clearly exacerbates their vulnerability to any mining-induced environmental degradation (Pegg 2006).

Consider the municipality of Governor Generoso in Davao Oriental with 45,000 people living on only 7,000 square kilometers of land suitable for agriculture (De la Cerna 2005). In Governor Generoso, 58 percent of the people live in poverty and the population overwhelmingly consists of subsistence farmers and subsistence fisher-folk. As Mayor Jerry de la Cerna stated, "We get our bread from the ocean and we get our bread from the land; we should protect the ocean and we should protect the land" (De la Cerna 2005, interview). Should an earthquake cause a tailings dam failure or should mine-pit dewatering aggravate an El Niño–induced drought, the environmental impacts of such an event would instantly thrust the poor from subsistence into destitution. In the context of such vulnerable populations, a hazard could clearly create a disaster. Consider as well the October 2005 tailings spills at the Rapu-Rapu Mine. After the spill there was substantial concern about whether fish caught in the Albay Gulf near Rapu-Rapu Island were safe to eat due to the presence of cyanide in the tailings and this concern caused the sales of fish caught in the Albay Gulf to fall to almost zero (Oxfam Australia 2008; Rapu-Rapu Fact Finding Commission 2006; Regis 2008). After the spill, fishermen had to leave their families to look for work in Manila and other nearby cities (Oxfam Australia 2008; Rapu-Rapu Fact Finding Commission 2006). Women found it necessary to find alternative employment that took them away from child-rearing duties while older children were recalled from school to help with household chores and to look after younger children. Some fisher-folk reported that they were forced to sell their fishing equipment in order to raise money; consequently, they would no longer be in a position to return to traditional fishing activities without assistance.

According to Father Ramoncito Segubiense,[8] director of the Social Action Center of the Diocese of Legazpi, the tailings spills caused up to 90 percent of the livelihood of the local people to be lost; those people who still fish in the area must now go progressively further and further out to sea in order to catch any fish (Segubiense 2009). Clearly, given the potential harm that may emanate from the interaction of a natural hazard with mining, and given the presence of large numbers of highly vulnerable people, the interaction of natural hazards with mining may create a disaster in the Philippines. In the words of Father Ramoncito Segubiense, "Mining could be a potential disaster for the community" (Segubiense 2009, interview).

Indigenous peoples: The most marginalized of the marginalized

In the Philippines, one group of people who are particularly vulnerable are the archipelago's indigenous peoples, those who have a historical continuity with the pre-Islamic and pre-Hispanic society of the archipelago (Holden 2005a; Holden and Ingelson 2007; Holden et al. 2011). When Islam was introduced to the archipelago in the fourteenth century, those who resisted Islam retreated to upland areas and continued their pre-Islamic animist belief systems. Then, after Spain colonized the Philippines, in the sixteenth century they retreated even further into upland areas to resist the Spanish. "The mountain communities around the country offered safe haven and a base for the resistance against colonialism" (Nadeau 2008, 30).

This historical process of indigenous retreat into mountainous areas has generated "one of the basic correlations in the Philippines," the fact that "indigenous peoples tend to occupy uplands since the lowlands were Islamized and Hispanized" (Rood 1998, 138). These people, who constitute approximately 15 percent of the population, primarily live in upland rural areas and engage in subsistence agriculture (Holden 2005a; Holden and Ingelson 2007; Holden et al. 2011). The largest concentration of indigenous peoples as a percentage of the population is in the Cordillera of the island of Luzon, where (collectively referred to as Igorots) approximately one-third of all indigenous peoples in the Philippines are found. The second largest concentration of indigenous peoples as a percentage of the population

---

8 The reader will note the numerous references to members of the Roman Catholic Church appearing throughout this book. The church is a profoundly important institution in Philippine society which since the Second Vatican Council 1962–65, moved away from the repressive tendencies of *La Frailocracia* and shown no shyness or reluctance to engage in activism on behalf of the poor and marginalized. This is discussed further in Chapter Six.

(but largest total number) is on the island of Mindanao (collectively referred to as Lumads), where approximately two-thirds of all indigenous peoples in the Philippines are found. The ratio of indigenous peoples to total population is lower on Mindanao due to that island's role as the traditional home of the archipelago's Muslims and due to the presence of large numbers of Christians who have migrated to Mindanao from other parts of the archipelago (Holden et al. 2011). There is also an appreciable concentration of indigenous peoples on the islands of Mindoro and Palawan; on the former, the indigenous people are the Mangyan and the Haunoo and on the latter, the indigenous people include the Palawan (or Palawano) of southern Palawan, the Tagbanua of central Palawan, and the Batak of northern Palawan (Holden et al. 2011).

One unfortunate characteristic shared by all indigenous people in the Philippines is poverty and marginalization. Their marginalization began in earnest during the American colonial period when their affairs were managed by the Bureau of Non-Christian Tribes (Leonen 2000). From 1901–13 the Bureau of Non-Christian Tribes was headed by Dean Worcester, the interior secretary of the Insular Government and a former zoology professor from the University of Michigan. Worcester described indigenous peoples as "the 'lowest of living men,' the first step in man's evolution from 'the gorilla and the orangutan'" (McCoy 2009, 101). The Bureau of Non-Christian Tribes was run by Worcester as "a bureaucratic empire that ruled nearly a million people, over 12 percent of the Philippine population" (McCoy 2009, 201). By 1908 Worcester would "proclaim himself 'the ruler of all non-Christians' before the hundreds of 'leaping and shouting savages' assembled for his annual inspection tours" (McCoy 2009, 211). Perhaps the classic quotation demonstrating the marginalization of indigenous peoples during the American colonial period was the famous definition of indigenous peoples provided by Justice Malcolm in *Rubi v. Provincial Board of Mindoro*, which described them as "natives of the Philippine Islands of a low grade of civilization, usually living in tribal relationships apart from settled communities."[9]

Father Albert Alejo, a Jesuit priest with a PhD in anthropology from the University of London, has stated that indigenous peoples "are on the lowest rung of society" (Alejo 2005, interview). Their political power is minimal and they suffer from inadequate human resource development (Alejo 2005). Rita Melecio, the Mindanao regional coordinator of Task Force Detainees of the Philippines (a human rights NGO created by the Association of Major Religious Superiors of the Philippines in 1974), echoed Alejo viewing indigenous peoples as highly marginalized and stating "whenever there are conflicts between indigenous peoples and other groups, the indigenous peoples

---

9 *Rubi v. Provincial Board of Mindoro*, 39 Phil. 693 (1939), as quoted by Leonen (2000, 25).

always end up being the losers" (Melecio 2005, interview). Anthony Badilla, the program coordinator of the Apostolic Vicariate of Palawan on the island of Palawan, has stated that indigenous peoples are the most marginalized in terms of education and basic social services (Badilla 2005).

The efforts of the Philippine government to attract mining investment have brought mining and indigenous peoples into dramatic contact with each other and the interaction of mining projects with indigenous peoples emerges from an intersection of geology with anthropology (Holden 2005a; Holden and Ingelson 2007; Holden et al. 2011). Mineral deposits, as explained in Chapter One, are usually found in mountainous regions because of the complicated orogenic forces that occasion their genesis. Figure 4.13 depicts the overlap between locations of mining projects with the percentage of each province consisting of land with a slope greater than 30 degrees. In the Philippines, indigenous peoples are also found in mountainous regions due to the historical process of indigenous retreat into mountainous areas and Figure 4.16 depicts the interaction of mining with indigenous peoples by overlaying the location of mining projects with the percentage of each province consisting of indigenous peoples.

It is estimated that half of all areas identified in mining applications in the archipelago are in areas inhabited by indigenous people (Holden 2005a; Holden and Ingelson 2007; Holden et al. 2011). Concerns about the environmental effects of mining are particularly acute among indigenous people due to the overlap of mineral resources with their ancestral domain and their reliance upon subsistence agriculture (Holden 2005a; Holden and Ingelson 2007; Holden et al. 2011). This overlap is most prominent in northern Luzon where, "Mining companies in indigenous communities in the six provinces of the Cordillera Administrative Region (CAR) have fouled rivers, endangered the environmental health of communities, and mistreated workers" (Foundation for Environmental Security and Sustainability 2007, 16). Abigail Bengwayan is the public information officer of the Cordillera Peoples Alliance (CPA). Bengwayan has stated that the spillage of mine wastes into the Abra River has led to the siltation of rice paddies and that this has poisoned waters and diminished rice production; this pollution has deprived many indigenous peoples of their livelihoods and contributed to poverty in the Cordillera (Bengwayan 2007). Concern about mining's environmental impacts also exists among indigenous people in other parts of the archipelago. Dionesia Banua is a member of the Tagbanua tribe, and the director of Natripal, an NGO, on the island of Palawan. Banua stated she is worried about a disaster happening on Palawan should mining continue there (Banua 2005). Intimately connected to the environmental effects of mining is a concern that mining may lead to the displacement of indigenous people. At the Taganito nickel laterite mine on the island of Mindanao, several families of the Mamanwa tribe displaced

Figure 4.16. Overlap between mines and indigenous peoples

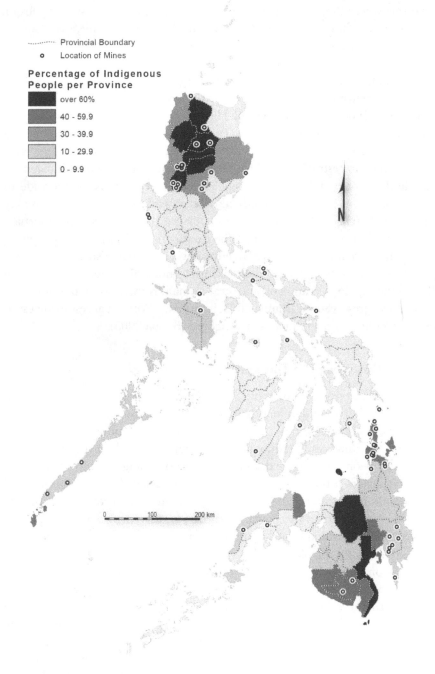

Source: Based on data from the Mines and Geosciences Bureau (2006, 2007a, 2007b, 2007c) and the United Nations Development Programme (2006).

by the Taganito nickel laterite mine, live beneath a bridge (Stavenhagen 2003). Once displaced by mining, indigenous peoples end up as poor urban migrants where they live in poor conditions lacking adequate shelter, jobs or basic services (Stavenhagen 2003). In Baguio City, for example, over half of the population consists of displaced Igorots, approximately 65 percent of whom suffer from extreme poverty (Stavenhagen 2003). Abigail Bengwayan stated that there have been several waves of Igorot migration into Baguio City that have occurred to escape poverty in the countryside (Bengwayan 2007). Magallanes Inocencio is the chair of Haribon Palawan, an NGO engaged in environmental and indigenous advocacy on Palawan. Inocencio regards mining as a threat to the cultural survival of indigenous peoples because it deprives them of their ancestral lands, which are the life support system for indigenous people (Inocencio 2005). Just as large areas of habitat must be set aside to protect biodiversity, large areas must also be set aside to protect ethnodiversity (Inocencio 2005). To Ajim Inni, an advocacy staff member with the Alternate Forum for Research in Mindanao, "The very identity of indigenous peoples is tied to the land; mining is a threat to ethnodiversity, it will displace indigenous people and cause them to lose their culture" (Inni 2005, interview). Dionesia Banua stated, "Preservation of ancestral domain is the most important thing for cultural preservation" (Banua 2005, interview). Without access to ancestral domain, cultural preservation is endangered (Banua 2005).

The Philippines: Too Dangerous for Mining?

The natural hazards present in the Philippines are all factors that, when taken into conjunction with the vulnerable population of the archipelago, could cause disasters even in the absence of any mining projects. Typhoons, earthquakes, tsunamis, volcanoes and El Niño–induced drought are all things that will impact vulnerable people and create disasters even if no mining were ever to occur in the archipelago. It is neither accurate nor fair to assert that these hazards will only occasion a disaster if they impact upon a mining project as they are fully capable of doing so on their own. Nevertheless, the environmental effects of mining (the high volumes of highly acidic and toxic tailings and its potential interference with hydrologic regimes) make it an activity that can exacerbate the risks presented by these natural hazards. Should one of these hazards impact upon a mine there will be a technological disaster as well as a natural disaster. To Father Ramoncito Segubiense, disasters can happen more often and be more severe if they are technological rather than natural; technological disasters are an even bigger threat than natural disasters (Segubiense 2009).

Given mining's potential for environmental harm, the natural hazards present in the islands and the dense populations of highly vulnerable poor people living on the brink of destitution, many civil society actors in the Philippines are of the view that the archipelago is too dangerous a place in which to locate large-scale mining projects (Holden 2005b; Holden 2011). Attorney Asis Perez, the senior staff lawyer for Tanggol Kalikasan (Defense of Nature, a public-interest environmental law office), regards the Philippines as perhaps the most dangerous place in the world to mine with its susceptibility to natural hazards, its small land area and high population (Perez 2005). Annabelle Plantilla, executive director of the Haribon Foundation (an environmental NGO), is of the view that the intrinsic vulnerability of the Philippines to natural hazards makes it an undesirable location for mining (Plantilla 2005). To Father Ramoncito Segubiense, the archipelago is too dangerous a location in which to locate large-scale mining projects and a Marcopper-style disaster could happen at the Rapu-Rapu Mine – he is worried that someday its tailings dam will completely fail, releasing massive quantities of toxic materials into the Albay Gulf (Segubiense 2009). Jesus Garganera stated that one of the starting points of the ATM in its campaign against mining is the argument that the Philippines are too dangerous a place in which to locate mining (Garganera 2009). Attorney Grizelda Mayo-Anda,[10] the founder and executive director of the Environmental Legal Assistance Center (ELAC) in Puerto Princesa City on the island of Palawan, holds the position that it is unjustifiable to allow mining to take place in the Philippines given its susceptibility to disasters (Mayo-Anda 2009).

Mining and the Bataan Nuclear Plant: Risky Activities in a Hazard-Prone Country?

An analogy may be drawn between the current efforts of the government to encourage mining and the ill-fated efforts of the government to develop the Bataan Nuclear Power Plant (BNPP) during the 1980s, as these are both discussions of hazardous activities being located in hazardous areas. During the early 1980s, the government of President Ferdinand Marcos paid USD 2 billion to the Westinghouse Corporation to build a nuclear power plant at Napot Point in Morong, Bataan (Figure 4.17), approximately 80 kilometers west of Metro Manila (Broad and Cavanagh 1993).

The nuclear power plant was planned to provide electricity for the Bataan Export Processing Zone but its site was highly controversial as it was near an earthquake fault line and was surrounded by several volcanoes (Broad and Cavanagh 1993). There was substantial concern that an earthquake could damage the cooling system in the plant's reactor and cause a loss of coolant

---

10 Vesilind (2002, 72) described Attorney Grizelda Mayo-Anda as a "dynamo" and ELAC as a "group of crusading lawyers."

Figure 4.17. The location of the Bataan Nuclear Power Plant

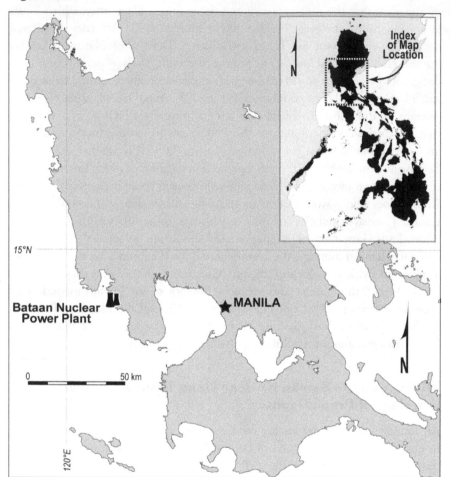

accident.[11] During the 1985–86 "snap election" campaign, Corazon Aquino promised she would not allow this "controversial and unsafe" plant to be operated and within days after the 26 April 1986 accident at Chernobyl, she carried out her promise and mothballed the plant (Broad and Cavanagh 1993, 123). Ricardio Saturay, at the CEC, definitely views the situation with

---

11 These concerns seem immanently reasonable given the loss-of-coolant accident which happened at the Fukushima Daiichi nuclear power plant in Japan on 11 March 2011. A magnitude 8.94 earthquake happened near the nuclear power plant and the reactors immediately went into emergency shut down procedure and were using diesel generators to circulate cooling water through their nuclear reactors. However, once a tsunami caused by the earthquake hit the nuclear power plant, the diesel generators were disabled and a serious loss-of-coolant accident occurred (British Broadcasting Corporation 2011).

the BNPP as analogous to that prevailing with mining in that they are both high risk activities located in hazardous areas (Saturay 2009). The CEC is currently involved in activism against plans to revive the BNPP and two of its major concerns are the geological hazards posed to the plant and the ability to dispose of nuclear waste safely in the Philippines given its propensity to geological hazards (Saturay 2009). Engineer Virgilio Perdigon is of the view that there is "a very close parallel between the BNPP and mining" (Perdigon 2009, interview). This nuclear power plant could have generated an incalculable amount of damage because of its proximity to Metro Manila and its location on top of an earthquake fault line (Perdigon 2009). To Dr Emelina Regis, mining amid these natural hazards is highly similar to the controversy surrounding the BNPP and just as an earthquake could have damaged the nuclear power plant causing a leak of radioactivity, an earthquake could destroy a tailings dam and cause a tailings spill (Regis 2009). Amalie Obusan, from Greenpeace Southeast Asia, regards mining as analogous to the BNPP; just as the BNPP was located on a fault line, many mining projects today are also located on fault lines and are also located in areas vulnerable to typhoons (Obusan 2009).

While the consequences of a loss of coolant accident at the BNPP would be infinitely greater than one tailings spill (such as the one occurring at the Rapu-Rapu Polymetallic Project in October 2005) there was only one BNPP and there are well over fifty large-scale hardrock mines being developed across the Philippines. Accordingly, a high magnitude earthquake, or super typhoon, which cases several tailings dam failures (or a widespread El Niño–induced drought that causes substantial diminution of groundwater resources) could collectively generate a swath of devastated ecosystems and impoverished communities rivaling the effects of a loss of coolant accident at the one nuclear power plant.

Given the overlay of mining with natural hazards and the assertion of many civil society actors that this renders the archipelago an unsafe location for a mining-based development paradigm, attention now turns to two things[12] that may arguably ameliorate the risks these hazards pose to mining: the subjection of mining projects to an environmental impact assessment and the use of technology. Perhaps subjecting mining projects to a thorough and rigorous environmental impact assessment could result in changes to the

---

12 The reader may wonder about using insurance as a device for ameliorating the risks these hazards pose to mining. In the Philippines, it is very difficult for people to acquire insurance protecting them from natural hazards (Lat 1994). Also, the vast majority of the population is unable to afford insurance. According to Bankoff (2003a, 61), the Philippines is "a country where private insurance remains the preserve of the few." Requiring mining companies to contribute to a calamity insurance fund will be discussed in Chapter Six.

location of mining projects away from vulnerable populations? Alternatively, perhaps the use of "state of the art" technology could provide tailings dams invulnerable to even the most powerful typhoons and earthquakes? Regardless of the specific measures used, can human ingenuity and technological skill overcome (possibly even tame) the forces of nature and make a mining-based development paradigm viable notwithstanding the natural hazards existing in the Philippines? These questions will be addressed in the next chapter as technocratic responses to the risks that are examined.

Chapter Five

TECHNOCRATIC
RESPONSES TO THE RISKS

Government and Industry Awareness of the Risks

The Philippine government and the mining industry are not oblivious to the risks presented by the intersection of mining and natural hazards and these risks are a concern to them. Mining investors are apprehensive that the government's "supportive policy could wane because of populist pressures in the event of a big mining disaster" (Landingin 2008, 9). Should a typhoon or earthquake collapse a tailings dam or if mine-pit dewatering aggravates an El Niño–induced drought, this could generate a backlash against mining that leads to an abrupt cessation of the government's support for mining. This would be an example of how "the enabling power of catastrophes" can "achieve and exceed the political significance of revolutions" (Beck 1992, 78). To both the government and the mining industry, these risks are relatively serious but are, however, quite capable of being overcome by the use of technocratic responses and the environmental effects of mining are more than capable of being managed.

Environmental Impact Assessment: A Tool of Environmental Management?

Introduction to environmental impact assessment

The first technocratic method of environmental management, which could be something capable of reducing the risks of a mining related disaster, is environmental impact assessment (EIA), "perhaps the most widely used tool of environmental management in the minerals sector" (Mining, Minerals, and Sustainable Development 2002, 248). According to Wood and Bailey (1994, 38), EIA may be defined as "a systematic procedure for considering the possible effects of a proposed action on the environment prior to a decision being taken on whether approval should be given for the action to proceed and, if so, how it should proceed." Environmental impact assessment is intended "to allow people to adjust development projects to enhance their benefits and to minimize their environmental costs" (Ross and Thompson 2002, 231). The National Environmental Policy Act of 1969, a United States federal statute, is generally regarded as the first legislated EIA process. It required all agencies of the United States government to prepare an environmental impact statement (EIS) documenting the impacts of major federal actions significantly affecting the quality of the human environment (Ortolano and Shepherd 1995).

Environmental impact assessment in the Philippines

The Philippines initially implemented its EIA system in 1977 when President Marcos issued Presidential Decree No. 1151, requiring an EIS to precede all actions, projects, or undertakings, which significantly affect the quality of the environment. In 1978, Presidential Decree No. 1151 was followed by Presidential Decree No. 1586, which authorized the Minister of Human Settlements[1] to name the lead agency responsible for undertaking the preparation of an EIS for "environmentally critical projects" and projects located in "environmentally critical areas." No project deemed to be either an "environmentally critical project," or a project located in an "environmentally critical area," could proceed without first submitting an EIS.[2] Upon a successful review of the EIS, the president (or his duly authorized representative) would issue an Environmental Clearance Certificate (ECC) and then, and only then, could the project proceed (Bravante and Holden 2009). In 1981, Presidential Proclamation No. 2146 provided definitions of those activities that would constitute environmentally critical projects[3] and what areas of the country would qualify as environmentally critical areas;[4]

---

1 The Ministry of Human Settlements was a "superministry" within the government headed by Marcos' wife Imelda. It was described by Thompson (1996, 52) as "the largest patronage machine in the country," which built schools, roads, recreational sites and ecology projects all displaying a prominent plaque bearing the name of Imelda Marcos.

2 Presidential Decree No.1586 required an EIS to be prepared for "environmentally critical projects" and projects located in "environmentally critical areas" but it did not provide definitions of these terms.

3 Heavy industries (nonferrous metals mining, iron and steel mills, petroleum and petrochemical industries, and smelting plants); resource extractive industries (major mining and quarrying projects, forestry projects, fisheries projects); and infrastructure projects (major dams, major power plants, major reclamation projects and major roads and bridges).

4 Areas declared by law as national parks, watershed reserves, wildlife preserves and sanctuaries; areas set aside as aesthetic potential tourist spots; areas which constitute the habitat for an endangered or threatened species of indigenous Philippine wildlife; areas of unique historic, archaeological, or scientific interests; areas which are traditionally occupied by cultural communities or tribes; areas frequently visited and/or hard-hit by natural calamities, geologic hazards, floods, typhoons, or volcanic activity; areas

major mining projects were specifically cited as being an example of an environmentally critical project and areas frequently visited or hard-hit by natural calamities, geologic hazards, floods, typhoons, or volcanic activity were specifically listed as environmentally critical areas. The system resulting from these presidential edicts formed the basis of EIA in the Philippines, and the Environmental Management Bureau (EMB) of the DENR became entrusted with implementing this system. It has had a substantial influence on it through the issuance of administrative orders (AOs), and a series of DENR AOs, culminating with DENR AO 2002–42 and DENR AO 2003–30, have determined the nature of contemporary EIA in the Philippines (Bravante and Holden 2009; Ingelson et al. 2009).

Environmental impact assessment of mining projects

The outline of the process

When a mining project proponent wishes to obtain an ECC it will consult the EMB. The EMB will have to decide whether the project is an environmentally critical project and whether the project is going to be located in an environmentally critical area. Generally, a large-scale mining project is almost always considered an environmentally critical project (Bravante and Holden 2009; Ingelson et al. 2009). After determining which EMB office will be responsible for the EIA, the EMB will then constitute an environmental impact assessment review committee that will then conduct a scoping exercise involving public hearings to determine the likely environmental impacts of the project. At the conclusion of the scoping exercise, the proponent will prepare and submit its EIS to the EMB and, at this point in time, the EMB is given 120 days to examine the EIS. The EMB often holds public hearings during the 120-day examination period to hear from concerned members of the public. The EMB may make written requests to the project proponent for additional information during this examination period but it is only allowed to make two requests for additional information and these requests may only be made during the first 90 days of the 120-day examination period. If the project proponent cannot comply with a request for information made by the EMB, the EMB is to make a decision based on whatever information is available so as to comply with the 120-day timeframe. At the end of the 120-day examination period the EIS will be deemed approved, unless

classified as prime agricultural lands; recharge areas of aquifers; water bodies tapped for domestic purposes, within the protected areas, or which support wildlife and fishery activities; mangrove areas; coral reefs.

expressly rejected, by the EMB and the ECC will be issued (Bravante and Holden 2009; Ingelson et al. 2009).

A process designed for rapid approval

These rules essentially impart a default period of 90 days into the process. If the EMB has not made its two requests for information within the first 90 days after having received the EIS submitted to it by the project proponent, the EMB will be unable to make any requests for information and will, in all likelihood, approve the project. To provide perspective on how short this time period is, consider that in the United States a large-scale hardrock mine on federal lands will routinely take somewhere between 18 months to 8 years to go through the EIA process under the National Environmental Policy Act (National Research Council 1999).

The importance of environmentally critical areas as spaces of vulnerability

Where the EIA system (at least in its 1981 incarnation under Presidential Proclamation No. 2146) offered potential as something capable of preventing a natural hazard induced mining disaster was its definition of areas frequently visited or hard-hit by natural calamities, geologic hazards, floods, typhoons or volcanic activity as "environmentally critical areas." This meant that, at least ostensibly, such areas were delineated as areas prone to natural hazards and that activities located in such areas would be subject to scrutiny. Conceivably, if such areas appeared to be too dangerous, activities capable of being adversely effected by natural hazards, such as mining, could be excluded from them. This appeared to go a long way towards recognizing how such areas are spaces of vulnerability where limitations should be placed upon what activities can and cannot occur. Unfortunately, however, subsequent DENR edicts have greatly reduced the role played by natural hazards in determining whether an area is an environmentally critical area. In 2002, DENR AO 2002–42 allowed the EMB to consult with the Department of Trade and Industry (DTI) in determining the technical definition of environmentally critical areas. This allows the EMB to take into account industrial and technological innovations and trends in determining what is, and is not, hazardous. Then in 2003, DENR AO 2003–30 removed hazard as a defining characteristic of an environmentally critical area. These changes demonstrate a downplaying of the risks posed by natural hazards to mining projects; the former shows how technology is something viewed as being capable of overcoming hazards and the latter shows how hazard vulnerability is not regarded as something capable of making an area environmentally critical.

Assessing the adequacy of the environmental impact assessment system

The use of the EIA system as a vehicle for reducing the risks of a mining related disaster is problematic. In general, the EIA system for mining projects in the Philippines has been viewed as "a tokenism, designed to make it appear as if mining projects are being evaluated with respect to their environmental impacts when really there is no serious intent to do so" (Bravante and Holden 2009, 524). "Mining interests," wrote Mayo-Anda (2000, 226), "are given undue support without properly balancing such egregious development with environmental protection." In the words of Bravante and Holden (2009, 540–1):

> The EIA system for mining projects in the Philippines is an example of what could be called 'going through the motions.' It is a process designed to make it appear that projects are being subjected to an environmental assessment while facilitating their inevitable approval; the priority of the Philippine government is the promotion of mining, not a reconciliation of mining's benefits with its possible environmental costs.

To Rovik Obanil, the EIA system should prevent harm from occurring when mines are located in areas vulnerable to natural hazards, but the implementation of the EIA system prevents this from happening as the system has been progressively changed in order to make it easier for mining projects to be approved (Obanil 2009). "In the Philippines, delay or abandonment of projects is never an option of the proponent" (Gatmaytan 1993, 27); as Attorney Grizelda Mayo-Anda stated, "this is EIA measured by a stopwatch and mandated by the skewed development priorities of the Philippines" (Mayo-Anda 2005, interview). Dr Emelina Regis views the provisions in DENR AO 2002–42 requiring the EMB to consult with the DTI in determining what is an environmentally critical area as something that allows the government to trade off the dangers mining poses to the environment in exchange for economic activity (Regis 2009). In the opinion of Engineer Virgilio Perdigon, if the EIA system were properly adhered to, mining would not be allowed in Bicol because of its vulnerability to typhoons (Perdigon 2009). Perdigon also objects to the EMB consulting with the DTI; the EMB should be concerned with protecting the environment instead of being concerned with encouraging economic activity (Perdigon 2009). The Rapu-Rapu Fact Finding Commission (2006) was highly critical of the EMB granting an ECC for the Rapu-Rapu Polymetallic Project when it did not take into account possible impacts upon the residents of the province of Sorsogon, only 12 kilometers across the Albay

Gulf from Rapu-Rapu Island. As the Rapu-Rapu Fact Finding Commission (2006, 95) wrote:

The EMB-DENR accepted and approved an Environmental Impact Statement (EIS) submitted by the company that failed to identify Sorsogon, just several kilometers across the bay from Rapu-Rapu Island, as one of the primary impact areas despite the obvious possibility that dispersion of mine waste materials could reach its shores when weather conditions favor the southward transport of the toxic plume.

Public participation in the environmental impact assessment process

The importance of participatory planning in disaster risk reduction

Arguably, the most serious problems with the EIA system as it applies to mining projects in the Philippines is its failure to provide meaningful opportunities for members of the public to participate in the EIA process. Participatory planning is often viewed as something essential to disaster risk reduction (Wisner et al. 2004). To Lyra Magalang, the Oxfam Great Britain program officer for disaster risk reduction in the Philippines, disaster prevention requires participatory planning; there must be participation by the effected community and everyone must know what to do (Magalang 2009). However just as public participation is required in order to prevent disasters, the EIA system has received substantial criticism for failing to avail meaningful opportunities for participation to members of the public (Bravante and Holden 2009; Ingelson et al. 2009; Yates 1993). As Hatcher (2010, 19) wrote, "In practice, the participation rights, including the right to information, participation in decision making and access to justice were found to be lacking."

The discretionary nature of public hearings

The first way in which the EIA system fails to provide meaningful public participation emanates from the discretionary nature of public hearings. Under the Philippine EIA system, public hearings are mandatory unless otherwise determined by the EMB (Bravante and Holden 2009; Ingelson et al. 2009). This means that public hearings are held at the discretion of the DENR and it may, if it wishes to, dispense with having them. Effective and meaningful public participation entails substantial time and expense. "The transaction costs – in time, labor, and expense – of responding to information requests, conducting public hearings or ensuring mechanisms for access to justice are often an annoyance to government agencies and businesses" (Bruch and Filbey 2002, 5). Given the express policy statement of President Macapagal Arroyo in Executive Order 270, to address the tedious permitting process for mines, one must wonder whether the mining promotion role of the DENR (as carried out by its MGB) has trumped its environmental protection role (as carried out by its EMB) and how willing the EMB will be to facilitate public hearings if these slow down the rate of mining investment in the archipelago. Vivoda (2008, 137) is firmly of the view that the provisions for public participation in the EIA system have been eroded in the name of "streamlining" the EIA process.

A narrow definition of who may participate

The second problem stems from a narrow definition of who may participate. "Stakeholders" are defined by DENR AO 2003–30 as those "who may be directly and significantly affected by the project or undertaking" (Bravante and Holden 2009, 536). This is a substantial departure from the broader definition found in its predecessor regulation, DENR AO 96–37, that stakeholders are "persons who may be significantly affected by the project or undertaking such as, but not limited to, members of the local community, industry, local government units, non-governmental organizations and peoples organizations" (Bravante and Holden 2009, 536). Restricting the definition of stakeholders renders other members of society who are affected (albeit in a less direct manner) "outsiders" who should not be part of the process. Lia Esquillo, executive director of Interface Development Interventions, an NGO in Davao City, indicated the project proponent routinely selects who attends the hearings and asks people to sign an "attendance list," which it then uses to prove that it has consulted affected communities (Esquillo 2007). There have also been documented instances where mining companies have asked people attending hearings to sign attendance lists and have later produced these same lists purporting them to be petitions in favor of mining (Hatcher 2010; Sanz 2007; Whitmore 2006). Father Medardo Salomia, the executive director of the Interfaith Movement for Peace, Empowerment and Development in Mati, Davao Oriental, stated that project proponents routinely only invite those in favor of mining (Salomia 2005). In the case of the Rapu-Rapu Polymetallic Project, the mining company assured that the "public consultation [would] be ongoing and integral to the Project" (McIlwain 2003, 94). Notwithstanding such an assurance, the public hearings were held on mining company property on Rapu-Rapu Island, participants were dependent on the project proponent for transportation to and from the island, people known to be opposed to the mine were not consulted and neither were any of the residents of the province of Sorsogon (Oxfam Australia 2008). The holding of the Rapu-Rapu Island hearings and the dependence of those who attended upon the mining company for transportation was a skillful use of geography as power by the mining company. Those who wanted to attend needed the mining company to take them to the mining company's property where the company controlled who was present and what was said. The Rapu-Rapu Fact Finding Commission (2006, 109) concluded, "These circumstances do not augur well for a real and meaningful participation absent a neutral ground."

Inadequate dissemination of information

The third problem with the public participation provisions of the Philippine EIA system is its failure to disclose information to members of the public. An essential component of disaster prevention is public information, "it should be accurate, simple and readily available" (Stenson 2006, 232). In many jurisdictions, information about pending projects is made available by making documents pertaining to that project available to the public (Bravante and Holden 2009).[5] In the Philippines, however, there is no required distribution of the EIS to the public as a way of soliciting public comment on the EIS (Bravante and Holden 2009; Ingelson et al. 2009). In fact, the Philippine environmental impact assessment process regards all environmental impact statements submitted by the mining project proponent to the government as confidential data; disclosure of such information rests on the discretion of the government (Bravante and Holden 2009; Ingelson et al. 2009). The 1987 Constitution of the Philippines[6] does enshrine a right to information on matters of public concern, but "no legislation sets procedures for access to information or penalties for officials who fail to disclose lawfully available data" (United States Department of State 2011, 18). Since "the government agency holds all the information secretly, the opposition cannot verify the EIS's underlying base analysis" (Yates 1993, 1147). To Attorney Grizelda Mayo-Anda, the lack of transparency is a challenge; the DENR must be pressured into providing documents and it is reluctant to do this voluntarily (Mayo-Anda 2005). According to Mayo-Anda, "there is an asymmetry of information in the EIA process" in that the mining project proponent and the government have more information about the project than do members of the public (Mayo-Anda 2005, interview). "Access to information is a fundamental requirement for participation, as it is a means (at least in part) of rectifying the imbalance of power between mining proponents and potentially impacted

---

5 For example, the state of Montana in the United States makes environmental impact statements of mining projects available at http://www.deq.mt.gov/eis.mcpx

6 Article III, Section 7.

people" (Goodland and Wicks 2008, 185). "It is hard for people affected by development plans to challenge those plans when developers and government agencies fail to supply them with detailed information" (Pye-Smith 1997, 27). In the Philippines, "Public scrutiny makes government agencies uncomfortable" (Yates 1993, 113). Any attempt at acquiring information about a mining project is regarded as a very challenging undertaking: "Looking for mining-related data, especially environmental clearances and pollution warnings issued to mining companies can be as difficult as digging up the precious metals themselves" (Aguilar 2008, 63). In the opinion of Attorney Ingrid Gorre, a staff lawyer with the Legal Rights and Natural Resources Center, there is a tremendous disparity in information between the community and the mining company (Gorre 2004). Attorney Emily Manuel, a colleague of Gorre at the Legal Rights and Natural Resources Center, indicated that the DENR is very reluctant to share information on any pending applications and will only disclose information about approved applications (Manuel 2004). "There is a complete and utter lack of transparency with respect to pending applications" (Manuel 2004, interview). Engineer Virgilio Perdigon articulated a view that there is a chronic lack of access to information regarding mining (Perdigon 2009). Lafayette Mining declared that it would "disclose relevant information early so that the key stakeholders are informed about developments before they occur" (McIlwain 2003, 94). Nevertheless, despite such an assurance of transparency, Perdigon has never been able to ascertain the answer to a question as rudimentary as whether the Rapu-Rapu Polymetallic Project is a cyanide vat leach operation or a cyanide heap leach operation because the mining company has never disclosed this (Perdigon 2009). Jesus Garganera indicated that a lack of access to information creates problems for the ATM; they simply do not know what types of technology mining companies are using and this lack of information makes it difficult for mining companies to have credibility (Garganera 2009). In the words of the Foundation for Environmental Security and Sustainability (2007, 38):

Communities often know little about the actual mining process and are poorly prepared to judge the nature and seriousness of accidents, real or alleged. Both government and mining companies do a poor job of communicating and sharing information with the public.

In Chapter Three, the Marcopper tailings spill was discussed. This spill generated a substantial amount of concern among the Philippine people about mining; the spill "reflected negatively on the minerals industry as a whole" (Maglambayan 2001, 78). The simple act of mineral exploration in many rural areas often generates consternation among people that they will be living near another Marcopper; "when a mining company just explores an area, people in the local communities already feel threatened" (Environmental Science for Social Change 1999, 79). If members of a community are denied information about the risks of a project near them, they are being exposed to a substantial degree of uncertainty; couple this lack of knowledge about a project with the given knowledge about what happened at Marcopper and opposition to mining becomes imminently understandable; as the Mining, Minerals, and Sustainable Development (2002, 293) report stated, "Secrecy does not build trust." In the words of Maglambayan (2001, 82), "the lack of accurate information about the activities of the minerals industry fosters distrust for the industry among the public." Indeed, Garb and Komarova (2001) even go so far as to state that withholding or manipulating of information about the environmental risks faced by people is a form of violence against them.

Where this lack of information becomes critically important is when certain actors in Philippine society are legally required to provide their consent to a mining project before it can proceed. Under the Local Government Code of 1991 (Republic Act 7160), the DENR must obtain the consent of the province, municipality and *barangay* (village) where a proposed mining project is to be undertaken (Holden and Jacobson 2006). Similarly, the Indigenous Peoples Rights Act (Republic Act 8371) requires the free prior and informed consent of all members of an indigenous cultural community who have their ancestral domain where a proposed mining project is to be undertaken (Holden and Ingelson 2007). If these local government units and indigenous communities are denied access to the EIS, they will be asked to consent to a project about which they lack full information, almost certainly a difficult thing to do. In the words of Orellana (2002, 236), "the links between prior informed consent and procedural rights are clear: without information and participation there can be no informed consent."

When copies of the EIS are made available to the public, they are long, complicated and highly technical documents written in English; often the EMB will translate the executive summary into a local dialect but usually some of its meaning will be lost during the translation (Bravante and Holden 2009). A frequent complaint from communities adjacent to mining operations is a lack of explanation of what is in EIA documents and the implications of the project for the adjacent communities (Hatcher 2010). As Attorney Ma Paz Luna (2005, 8), director of the environmental NGO Tanggol Kalikasan wrote:

A critical task that needs to be undertaken in order for participation to be meaningful is the use of plain and understandable language in the EIS.

Aside from the fact that, in practice, the EIS frequently uses English; many are compounded with highly technical language, tables and graphs that hardly translate to issues that local communities would readily grasp. Participation would be cosmetic and token if there is no real appreciation by the community of the contents of an EIS.

Arnstein's hierarchy of citizen participation

According to director Emmanuel Isip, the quality of provisions for public participation in the Philippine EIA system receives a rating of 5 on a scale from 1 to 10 (Isip 2005). To Isip, many view the public participation provisions in the EIA system as an "exercise in futility" (Isip 2005, interview). The major problems with the public participation provisions of the EIA system are outlined in Table 5.1.

A conceptual device frequently used in discussions of public participation is the hierarchy of citizen participation developed by Arnstein (1969), and presented in a seminal article in the *Journal of the American Institute of Planners*. Arnstein, in explaining the significant gradations of citizen participation, stipulated eight steps through which public participation can pass through and these steps are displayed in Figure 5.1.

Table 5.1. Issues in the public participation provisions of the EIA system

Issues	What this entails
Discretionary nature of public hearings	The discretion to hold public hearings lies with the government and there is no mandatory requirement for them to be held.
Narrow definition of who may participate	"Stakeholders" are defined by the government as those "who may be directly and significantly affected by the project." This makes other members of society who are affected in an indirect manner "outsiders" who should not be part of the process.
Transparency versus confidentiality	Environmental impact documents are not required to be distributed to the public. In fact, they are considered confidential. This creates an asymmetry of information between the proponent and the government (on one hand) and the public (on the other hand).
Complicated and highly technical documents	This makes the process beyond the comprehension of most members of the public and renders it cosmetic and token.

Figure 5.1. Hierarchy of citizen participation

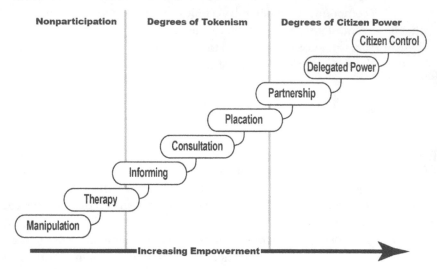

Source: Adapted from Arnstein (1969).

The bottom steps, "manipulation" and "therapy," are degrees of nonparticipation. At these levels, participation is nothing more than a public relations exercise designed to gain support for a predetermined decision. Whatever participation takes place here is conducted for the purpose of educating the public and this is a purely manipulative process – not a true form of participation. The middle steps, "informing" (where there are flows of information from managers to citizens but not dialogue), "consultation" (where citizens are given a voice but are not necessarily heard), and "placation" (where managers seek citizen advice but retain decision-making authority), are degrees of tokenism. At these levels, participation has evolved to become a consultative process where citizens may be heard but not necessarily heeded by decision makers. Finally, the highest steps, "partnership," "delegated power" and "citizen control," represent the highest degrees of decision-making clout. The public participation in the Philippine EIA system is – in the hierarchy of citizen participation developed by Arnstein (1969) – within the degrees of tokenism, namely a consultation. At the level of consultation, a step towards full participation has been taken but Arnstein (1969, 219) argues that it "is still a sham since it offers no assurance that citizen concerns and ideas will be taken into account." Here, "participation is measured by how many [people] come to meetings, take brochures home, or answer a questionnaire" (Arnstein 1969, 219). Mayo-Anda is of the view that the EIA system is a consultation rather than a genuine attempt to encourage public participation (Mayo-Anda 2005). The process is more of an information dissemination process than a participation method and there is no way it can influence the decision-making procedure (Mayo-Anda, 2005). Attorney Augusto Gatmaytan, a lawyer and anthropologist based in Davao City, stated that "it is one thing to be consulted, it is another thing to be listened to; the government is going through the motions of having public participation" (Gatmaytan 2005, interview). The system is arranged to make it look as if the mining project has widespread support when it does not provide enough information to the public regarding mining. To Arnstein (1969), public hearings are one of the most frequently used methods for engaging in consultation. Their use as a minimal degree of tokenism is made more evident in the Philippine EIA system by their discretionary nature. The government is under no legal obligation to hold them, there is a highly constrained definition of who may participate in them and there is no required distribution of the EIS to the public as a way of soliciting public comment on the EIS. With such limited facilities for public participation, it is unlikely that the EIA system can become the participatory planning process necessary for disaster risk reduction.

A denial of environmental justice

A clear argument can be made that these limited facilities for public participation constitute a denial of environmental justice, which can be defined as "the right to a safe, healthy, productive and sustainable environment for all" (Draper and Mitchell 2001, 93). In recent years, as marginalized communities around the world have confronted various types of challenges to the environment upon which they depend for their livelihoods, the environmental justice movement has emerged. This movement takes environmental issues and reiterates them from a social justice perspective taking into account the needs of the poor and marginalized. An essential tenet of environmental justice is that people should be entitled to meaningful participation in the decisions that affect them (Kuhn 1999). If a community near a proposed mining project cannot participate in the environmental assessment of that project, it is being denied environmental justice. As Kuhn (1999, 648) wrote:

Increased public participation is essential to achieving environmental justice. True public participation and environmental justice cannot be realized until the communities that are impacted by environmental regulations have a voice in the process equal to that of regulated industry.

Technology as the Solution to the Risks

Technology can prevent disasters

The second technocratic method of environmental management, which could be something capable of reducing the risks of a mining related disaster (and one that often follows as a condition of EIA approval),[7] is the use of technology. Nelson Devandera, executive director of the Palawan Council for Sustainable Development,[8] stated he is confident that technology can prevent any disasters (Devandera 2005). "Mining companies can spend in the name of the environment; they operate for profit and for honor" (Devandera 2005, interview).

Technological responses: A controversial topic

The pollution haven hypothesis

Technological responses to the challenge of mining related environmental disruptions are a highly controversial topic as there are many writers who allege that the mining industry is moving from the developed world to the developing world to escape the imposition of strict environmental protection standards in the former (Brown et al. 1999; Dobb 2002; Evans et al. 2001; Gedicks 2001; Muradian et al. 2003; Sampat 2003; Veiga et al. 2001).[9] As Muradian et al. (2003, 776) wrote, "Rising difficulties for exploiting public lands, and growing environmental concerns seem to have played an important role in 'pushing back' mining activities from the industrialized world." According to Veiga et al. (2001, 193), "Although multinational firms often claim that they apply the same environmental practices in less-developed host countries as in their country of origin, this may not always be the case." In the words of Sampat (2003, 114–15):

> Multinational mining companies have increasingly focused their quest in the developing world, where mines can be worked more cheaply as

---

7 The ECC for the Rapu-Rapu Polymetallic Project required, as a condition of approval, that the proponent must use "adequate safety gadgets."

8 The Palawan Council for Sustainable Development is the body created by the Strategic Environmental Plan for Palawan Act (Republic Act 7611) and entrusted to implement environmental governance on the island of Palawan, which is, effectively, a giant protected area.

9 The differences in time frames for the environmental assessment of a large-scale hardrock mine in the Philippines (essentially 90 days) and the United States (between 18 months and 8 years) add credence to this view.

labor costs are lower and environmental regulations are typically not as strict as in Australia, Western Europe, or North America.

Best practices in environmental management

There are, however, other writers who dispute the allegation that mining companies seek pollution havens (Auty and Warhurst 1993; Warhurst 1992, 1999). To these writers, the mining industry has made a commitment to the implementation of "best practices in environmental management," which can be defined as "the methods and techniques that have proved to lead to successful outcomes through their application... most major mining companies are committed to the continuous improvement of their environmental and social performance" (Mining, Minerals, and Sustainable Development 2002, 249). Nelia Halcon, the executive vice president of the Chamber of Mines of the Philippines, is steadfast that the safety and environmental records of mining companies "have improved over the years because mining companies have internalized environmental management systems" (Landingin and Aguilar 2008, 18). As Warhurst (1999, 23) wrote:

> Dynamic companies are not closing down, reinvesting elsewhere, or exporting pollution to developing countries with less-restrictive regulatory regimes. Rather, these companies are adapting to environmental regulatory pressures by innovating, improving, and commercializing their environmental technology and environmental management practices, at home and abroad.

Industry reluctance to bear the necessary costs

In many ways, technology may well be the solution to the problems posed by mining amid natural hazards. In the case of tailings dam stability, there is no technical reason why a tailings dam cannot be built that is impervious to a direct hit from a typhoon or a high magnitude earthquake. Indeed, it has long been known that there is no technical reason why mines cannot be built resilient enough to with stand natural hazards such as those found in the Philippines; in the case of earthquakes, for example, "it is possible to design and construct aseismic tailings dams following the usual procedures for earth dam construction" (Dobry and Alvarez 1967, 259). However, constructing such a veritable fortress of a tailings dam will require a substantial outlay of expenditure. As Vick et al. (1985, 922) wrote:

> Insofar as mining is an intrinsically risky enterprise from an economic standpoint, mining decision-makers are accustomed to trading off

acceptable economic risks for potential benefits, and recognize that extremely low failure risks can be purchased only at an extremely high initial cost.

With respect to the hazards imposed by El Niño–induced drought, mining companies insist that their water use is highly efficient. As Xstrata (2008, 30) the project proponent of the Tampakan Project stated:

> A detailed surface water model is being developed in current feasibility studies. Modeling and catchment analysis work is currently being reviewed to include risks associated with production demand, downstream baseline flows against various criteria including that of one in 1,000-year drought events. This work will continue in 2009 as part of the feasibility and Environmental and Social Impact Assessment program. The water management plan will be developed in the context of both climate change and water stress and factor in the demands of the social population and environment.[10]

If technology can transcend the risks posed to mining projects by the hazards prevalent in the archipelago, then the first problem to emerge from such a technocratic solution is whether the mining industry will be willing to employ such technologies in the Philippine context. Engineer Virgilio Perdigon strongly doubts that a mining company will build a tailings dam with an extremely low rate of failure in the Bicol Region because doing so would entail operating at a loss (Perdigon 2009). Bicol is so vulnerable to typhoons and earthquakes that constructing a tailings dam robust enough to withstand such hazards would be extraordinarily expensive (Perdigon 2009). Jesus Garganera echoes Perdigon and views the acquisition of extremely low failure rates through technology as something that would be very expensive in the archipelago (Garganera 2009). To Ricardio Saturay, investing more in safety measures reduces risks but it also increases costs, therefore reducing profits (Saturay 2009). As Whitmore (2006, 311) wrote, "The mining industry is exceedingly quick to herald beneficial improvements in technology. However, history has shown us that

---

10 The authors find Xstrata's assurance that it will use a 1,000-year drought event as a baseline highly problematic. First, Xstrata cannot obtain accurate precipitation data going back to the year 1008 as no records were kept going back that far on the island of Mindanao. Second, accurately predicting precipitation 1,000 years into the future will be extremely difficult to do given uncertainty surrounding the magnitude of climate change occurring over the next 1,000 years (Yumul et al. 2011). Any such predictions will be probabilistic models, at best, and certainly not a robust statistical model designed with a high degree of reliability.

even if beneficial, these technologies are not necessarily used, as costs can sometimes be prohibitive." The logic of capitalism requires cost cutting to maximize profits (Emel and Krueger 2003).

Government reluctance to mandate the appropriate technology

Some stakeholders have articulated a concern that even though technology may overcome the archipelago's set of natural hazards, the government of the Philippines is unwilling to require it to be implemented. Rovik Obanil stated that low failure rates can only be purchased at an extremely high initial cost; in the Philippines, mining companies simply view such costs as unacceptably high because the risks are not borne by them due to the lax regulatory regime prevailing there (Obanil 2009). If a mining project proponent fails to comply with whatever legal requirements are placed upon it, there is a low probability that it will ever be convicted and even if it is convicted, the fines imposed upon it will be minimal (Obanil 2009). There is evidence that such concerns are well founded; from 1985 to 2003, there was nearly one mining-related pollution incident reported every year in the Philippines (World Resources Institute 2003). In almost 40 percent of these cases, no fines were imposed and when fines were imposed they were less than USD 5,000 (World Resources Institute 2003). Vivoda (2008, 140) adopts the view that the "government has allowed its reputation for not strictly enforcing environmental standards or protecting indigenous rights to act as an unstated selling point to potential investors." The 2010/2011 Fraser Institute survey of mining firms supports this claim because 15 percent of the 494 respondents to that survey cited environmental regulations in the Philippines as a factor that encourages investment in the archipelago while only nine percent stated that they would not invest in the Philippines due to its environmental regulations (Fraser Institute 2011). To Dr Emelina Regis, the government has adopted a view that the benefits of mining outweigh its risks and the government is unwilling to impose strict standards on the mining industry because mining is perceived to be a source of economic benefit. (Regis 2009). As Vivoda (2008, 140) wrote: "Many in mine-affected communities believe that the Philippine government has made an implicit commitment not to enforce costly social and environmental minimum standards justifying the costs to people and the environment by focusing on the long-term gains."

Technology is only as good as the people who use it

Ultimately, technology may not be enough to overcome such serious natural hazards. To Ricardio Saturay, technology – in theory – can allow a mining project to be constructed in such a way as to be capable of withstanding natural hazards but this would require adequate baseline data, much of which is unavailable at a local level (Saturay 2009). Also, technology may be state of the art but it is only as good as the people who use it, as "technology is commanded by people" (Saturay 2009, interview). The first of the two tailings spills at the Rapu-Rapu Polymetallic Project in October 2005 was caused by a plastic water bottle getting into pumping equipment and causing it to clog, thus generating an overtopping of the tailings dam (Oxfam Australia 2008). In this case, someone had not been diligent in ensuring that debris was kept out of the tailings dam and a system failure occurred. Rovik Obanil does not view technology as being something capable of preventing a large-scale disaster; perhaps the best example of this was Marikina City, the area of Metro Manila that was severely flooded by Typhoon Ondoy in September of 2009 (Obanil 2009). In October of 2008, less than one year before Typhoon Ondoy, the Asian Development Bank issued a report touting the flood control measures in Marikina City, and declared it "Flood Ready Marikina City" (Obanil 2009). Nevertheless, Marikina City was severely flooded and images of its residents attempting to cope with the floodwaters were displayed worldwide by the global media.[11] After the tailings spills at the Rapu-Rapu Polymetallic Project, President Macapagal-Arroyo constituted a fact-finding commission to investigate the causes of the tailings spills and the Rapu-Rapu Fact Finding Commission (2006, 88) concluded that "typhoons, heavy rainfall most times of the year and steep topography are factors that cannot be mitigated." Whitmore (2006, 312) adds: "If technical fixes are the answer to everything then we should have seen reductions in the number of tailings dam problems," with Bebbington et al. (2008, 908) urging that "while mining companies insist that their water use is highly efficient, communities and activists remain unconvinced and hydrologists tell us that the effects of removing large parts of rock in headwater areas can have non-linear, negative effects on water availability downstream." Father Ramoncito Segubiense does not view the mining industry as having the skills necessary to protect a community ten, or twenty, years into the future and wants an assurance of a continuous livelihood for the community in perpetuity; to Father Ramoncito, this is something the mining industry is unable to provide (Segubiense 2009). In the words of Mayor Jerry de la Cerna, "technology can help but technology cannot contain all the wastes forever" (De la Cerna 2005, interview).

---

11 See the photomontage prepared by the British Broadcasting Corporation (2009) entitled "In pictures: Philippines floods" (http://news.bbc.co.uk/1/hi/8276374.stm – accessed 10 November 2011).

Australian and Canadian mining companies operating in an unfamiliar environment

Concerns have also been raised with respect to the ability of Australian and Canadian mining companies to operate mines in the Philippines. Australia and Canada are, as pointed out in Chapter Two, the home countries of many of the mining companies operating across the archipelago. However, they are both countries with low population densities (Figure 5.2), which are tectonically stable. Canada is completely free from typhoons and Australia only experiences tropical cyclones in its most northerly latitudes.

A mining company may have substantial experience developing mines in the outback of Australia or the boreal forest of northern Canada where there are minimal natural hazards and few, if any, vulnerable populations capable of being adversely affected by mining. This mining company may then attempt to develop a mine in the high hazard setting of the Philippines and suddenly find that what worked at home may no longer work in the archipelago. If Canadian mining companies are unable to prevent beavers from disrupting their tailings dams one may reasonably wonder how they will cope when a typhoon, containing the energy equivalent of an atomic bomb, impacts their mine. Given the difficulty Canadian mining companies have shown coping with the threats posed to their tailings dams by semiaquatic rodents, it is not surprising that Seck (1999, 139) wrote, "Canadian mining companies are becoming well known

Figure 5.2. Persons per square kilometer, Australia, Canada and the Philippines (2011)

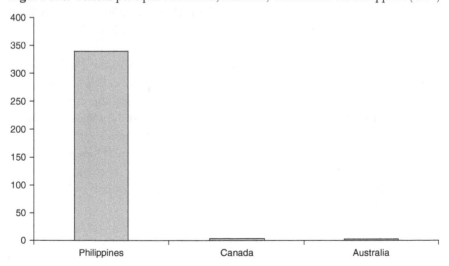

Source: United States Census Bureau (2011).

for the environmental disasters that all too often flow from their activities abroad, particularly in the developing world."

Mining companies often use the same technologies in their overseas operations that they use in their home countries (Hilson 2000). This is a point frequently made by the mining industry to refute allegations that it is moving to the developing world to escape stringent environmental regulations in the developed world (McMahon et al. 1999; Warhurst 1992, 1999). In Latin America, for example, "many of the treatment and control measures that have been installed are replicas of the site-specific models designed for use at North American sites" (Hilson 2000, 79). Writing in a World Bank Technical Paper, McMahon et al. (1999, 1) confidently stated, "new investment in large and medium sized mines often follow those standards more strenuous than those of their host country, given that their technology is based on the standards of their home country." This can, however, create problems when environmental conditions in foreign countries differ from those found at home. In Romania, for example, concerns have been raised about Australian mining companies implementing environmental management regimes designed for the arid environment of the Australian outback that are inappropriate when used in the higher rainfall environment of Romania (Mara and Vlad 2009). Similarly, concerns have been raised about Australian and Canadian mining companies importing their own environmental management regimes into the Philippines. Engineer Virgilio Perdigon is of the view that when foreign mining companies develop mines in the Philippines, they use technology that has been developed in their home countries and, in doing so, engage in technological ethnocentrism (Perdigon 2009). As Annabelle Plantilla (2006), the executive director of the Haribon Foundation, stated:

> Imported technology may be very difficult to adopt in the Philippines. First, our country is archipelagic and very mountainous. It is found near the equator, thus we have tropical rainforests. Most of the target areas for mining have people living in them. Conditions in the Philippines are very different from the temperate and continental countries of Australia or Canada where most of the technologies come from. These countries are very large tracts of land, which may be devoid of communities and where mining footprints are incomparable to those that will be left in an archipelagic country like ours.

The October 2005 tailings spills at the Rapu-Rapu Polymetallic Project provide an excellent example of a foreign mining company being unprepared for the environmental conditions prevalent in the Philippines. Andrew McIlwain, the chief executive officer of Lafayette Mining Limited (the Australian corporation that was a part owner of Lafayette Philippines Incorporated), declared that Lafayette had made commitments to best practices in environmental management and that these principles underpinned all of its operations. It declared that "its large structures, such as the open pit, tailings dam and waste dump [had] been designed to take into account environmental risk issues" (McIlwain 2003. 91). It also stated that "particular attention [had] been paid to both design of facilities and standard operating procedures, to ensure that chemicals and plant streams will not introduce spillage into the local environment... Drainage provision [had] been specifically made for the plant site to ensure that facilities [could] continue to run and [were] not compromised during periods of high rainfall" (McIlwain 2003, 92). McIlwain (2003, 99) then asserted:

> Lafayette believes that it has identified all aspects of the project where material risk could arise and has, either through design, contract or policy, developed appropriate response strategies which have the effect of either eliminating the risk totally or reducing the likelihood of occurrence to negligible instances.

Nevertheless, despite these confident assurances, this mine was responsible for two cyanide-laced tailings spills during its first month of operation! The second tailings spill at the Rapu-Rapu Polymetallic Project resulted from the Australian mining company lacking familiarity with the high precipitation levels prevalent in the Bicol Region (Saturay 2009). Specifically, the proponent built the tailings dam incrementally, making it progressively higher as more mine wastes were produced. The thinking behind such a method is that early in the life of the mine, when a small amount of waste had been generated, only a small tailings dam would be needed; later in the life of the mine, after more wastes had been generated, a larger dam would be needed. This may be an acceptable technique in the low precipitation environment of the Australian outback but in the Bicol Region – with its propensity for heavy rainfall events – a tailings dam should be built to accommodate the maximum amount of rain that could fall, not the amount of tailings generated by production at the mine. The mine was very early in its lifetime when a heavy rainfall event occurred, and this resulted in an overtopping of the dam and a spill of toxic materials. The Rapu-Rapu Fact Finding Commission (2006) insisted that acid mine drainage mitigation measures should not be based on best practices in other countries but should instead be based on the particular geophysical conditions of the Philippines. Attorney Grizelda Mayo-Anda opined that Australian and Canadian mining companies are not prepared to mine in the Philippines as their large landmasses are too different because of geography, geology and biodiversity (Mayo-Anda 2009). Australian and Canadian mining companies as lack the skills and experience necessary to mine in the conditions they will encounter in the Philippines (Obanil 2009).

Weak Governance of Mining in the Philippines

The concept of governance

Given the potential for environmental harm inherent in mining and given the problems that may emerge from locating mines in areas susceptible to natural hazards and inhabited by vulnerable populations, it is imperative for mining to be subjected to strong governance. If strong governance of mining is not available, there will be an increased risk of a mining related disaster. As the World Resources Institute (2003, 36) wrote, "Proposed mines in areas with multiple hazards and vulnerabilities may be especially problematic as they imply the need for careful and deliberate decision making that may be less likely in areas of weak governance." There are many in the Philippines who believe that, in the absence of good governance and respect for the rule of law, mining (particularly on the scale prompted by the government) will have serious long-term environmental costs (Doyle 2009). As Vivoda (2008, 136) wrote, "A weak Philippine state has produced weak and inefficient institutions of governance."

What exactly is the concept of governance? According to Labonne (1999, 316) "governance" is a broad process wherein "society manages its economic, social, and political resources and institutions for the development, integration, social cohesion and well-being of the people... [governance] encompasses the overlapping of three domains: state, civil society and the private sector." There is a clear distinction between governance and government. Governance refers to "the structures and processes that enable governmental and nongovernmental actors to coordinate their interdependent needs and interests through the making and implementation of policies in the absence of a unifying political authority" (Krahmann 2003, 331). Governance is a fragmented concept that is spread among both governmental and nongovernmental actors. In contrast to governance, government is defined as "policymaking arrangements and processes that centralize political authority within the state and its agencies" (Krahmann 2003, 331). Since governance involves a fragmentation of political authority among governmental and nongovernmental actors, there can be a pitting of governance against government with the latter being understood as the centralization of authority within the state (Holden and Jacobson 2006).

In the Philippines, the government – centralized within the state – is attempting, through its policymaking arrangements and processes, to use mining as a vehicle for encouraging the acceleration of the economy. However, the Marcopper tailings spill incident has generated a fear of mining among many communities adjacent to mining projects. In contrast to the government is governance, a fragmented authority spread among both governmental and nongovernmental actors. The opposition of these people to the government's efforts to encourage mining is, therefore, a demonstration of governance being posed against government. Governance exists apart from government and is spread among a number of actors. Governance is a concept of which the mining industry must increasingly be aware. It is not enough for a project proponent to have the approval of the government; rather, the project proponent must have the approval of those residing where the mine is to be located. If a mining project proponent does not address the wider issues of governance actors and tries to develop its project with only the approval of the government, it can expect to encounter opposition to its project. In the Philippines, however, there are a large number of factors unique to the archipelago's particular historical and spatial specificities that generate a substantial weakening of governance. The government is strong and is determined to implement its pro-mining agenda, but these factors (which weaken governance) substantially increase the risk of a mining related disaster.

The capture of the state by powerful forces

The first factor contributing to weak governance in the Philippines is the capture of the state by powerful forces. In Chapter One the historical origins of the Philippine oligarchy were discussed. This oligarchy has come to take control of the state and use it as a vehicle for advancing its class interests. "The state," wrote Hawes (1987, 53), "has been an instrument for class domination." Since this powerful ruling class controls the state and also stands to benefit through its partial ownership (along with foreign investors) of many mining firms, this class is reluctant to impose – and implement – strict rules and regulations upon the mining industry. The mining industry is reluctant to design tailings dams that are robust enough to withstand typhoons or earthquakes because such tailings dams would be inordinately expensive and the logic of capitalism requires cost cutting to maximize profits. In the archipelago, one characteristic of the state stands above all others: "the state defends the general interest of capital" (Hawes 1987, 53). The Philippines is governed by "a parasitic leader class that [uses] office for enrichment" (Constantino 1975, 402). The Philippine government is controlled by a "ruling class far more concerned about their intertwining networks of family and friends rather than the needs of a people in distress" (Kirk 2005, 20). "The basic institutions and governance structures in the Philippines are dominated by powerful vested interests, who, through informal influences such as patronage, politics, corruption, cronyism and clientelism, control the Philippines state"

(Vivoda 2008, 137). With vested interests having control of the state, the DENR is "seen as ineffective and subject to capture by powerful political and economic interests" (Stark et al. 2006, 1).

A lack of state resources to regulate mining properly

The second factor present in the Philippines imparting difficulties into the governance of mining is a lack of state resources to regulate mining properly. The government of the Philippines has been promoting mining so aggressively, concerns have been raised that this is placing "unprecedented pressures on [the] regulatory and oversight mechanisms of the mining sector" (Foundation for Environmental Security and Sustainability 2007, 14). According to Vivoda (2008, 136), "the government lacks the resources to enforce rules across the mining sector." Oxfam Australia (2008, 39) contributes "There is strong criticism in the Philippines of the capacity of government authorities to provide effective regulatory oversight of mining company activities." The DENR lacks the autonomy "required to effectively enforce the overall social and environmental safeguards enshrined in the mining regime" (Hatcher 2010, 13).

At the exact same time that the government has embarked upon an aggressive promotion of mining, it has actually experienced a reduction in the number of personnel employed by the DENR to regulate mining. The MGB, according to Landingin and Aguilar (2008, 19), consistently loses "skilled personnel to industry while being prevented from hiring more people because of limited budgets and a government wide [hiring freeze] policy." The DENR has few experts on natural resources management as most have left to become private sector consultants (Vivoda 2008). Salaries in the MGB are approximately 80 percent lower than in the mining industry so the MGB is constantly losing talented employees to them (Landingin and Aguilar 2008). In fact, from 2007 to 2008, the MGB lost 5 percent of its permanent workforce while also seeing a 99 percent increase in unfilled positions over the same time period (Landingin and Aguilar 2008). To Dr Emelina Regis, the provisions in the ECC for the Rapu-Rapu Polymetallic Project requiring the project proponent to conduct monthly pollution monitoring checks and to submit the results of those checks to the DENR indicate the inability of the government to regulate mining; the government, lacking the resources to monitor the mine, has entrusted environmental monitoring to the mining company itself (Regis 2009). The Rapu-Rapu Fact Finding Commission pointed to a lack of regulatory oversight by the DENR as a major factor contributing to the two tailings spills in October 2005. Government monitoring of the mining company "was not to best practice standards. It lacked the rigorousness and strictness to properly police an environmentally critical operation such as mining" (Rapu-Rapu Fact Finding Commission 2006, 94). Indeed, the Rapu-Rapu Fact Finding Commission (2006, 95) went so far as to declare "The DENR has lost its credibility as a regulating agency of mining." As Hatcher (2010, 13) wrote:

> While the chronic issue of the flight of experts to the private sector goes a long way to explain the shortage of qualified staff within government ranks, the Department appears to lack the very resources to carry out its socio-environmental mandate.

Conflicts of interest at the Department of Environment and Natural Resources

The third factor contributing to a weak governance of mining is the perception that the DENR is riddled with conflicts of interest. Through its EMB, the DENR is responsible for regulating the environmental management of mines. However at exactly the same time through its MGB, the DENR is responsible for the promotion of mining as an economic activity (Maglambayan 2001). There is a dual role assigned to the DENR where it is expected to act as both a promoter and regulator of the mining sector (Hatcher 2010). In the words of Stark et al. (2006, 7), "the DENR is given the conflicting dual role of promoting, mining and safeguarding the integrity of the environment and natural resources." As Bello et al. (2009, 236) wrote:

> On the one hand, the DENR is supposed to conserve the environment and control the extraction of natural resources. And yet, on the other, it is responsible for granting the permits to corporations from which the agency is supposed to protect the environment.

Both Michael Cabalda and Director Isip downplayed the dual role played by the DENR as both regulator and promoter of mining. In the opinion of Director Isip, there are no conflicts arising from the EMB and the MGB simultaneously being bureaus of the DENR (Isip 2005). To Cabalda, the MGB has the expertise to regulate mining and it is good to both promote and regulate; besides, if you want to regulate an industry you must have a good understanding of that industry (Cabalda 2004). Nevertheless, notwithstanding the assurances of these DENR employees, this "dual function" of the DENR has generated controversy. Maglambayan (2001, 84) recommends that the MGB "be transferred from the DENR to either the Department of Trade and

Industry, the Department of Science or to the Department of Energy." The present structure of the DENR has been described as being "schizophrenic" (Bello et al. 2009, 236). The Rapu-Rapu Fact Finding Commission (2006, 96) noted this conflict of interest stating, "For the MGB and EMB to be able to efficiently provide the services they are mandated to do, these must be totally independent from one another, which is precluded by the current set-up of the DENR."

The high levels of corruption prevailing in the Philippines

Corruption in the Philippines

The fourth factor contributing to a weak governance of mining is the high levels of corruption prevailing in the Philippines. The extent of corruption in the archipelago is deeply embedded and well documented (Bello et al. 2009). "Corruption," wrote Bankoff (2003a, 100), "is symptomatic of a 'culture' that permeates all levels of the public service." Kirk (2005, 3) described the society prevalent in the archipelago as one "in which nepotism, bribery, gift-giving and exchange of favors are the rule not the exception." President Macapagal-Arroyo was "the daughter of Diosdado Macapagal, the corrupt ex-president notorious for his government's involvement in smuggling" (Kirk 2005, 34). Macapagal-Arroyo spent much of her presidency embroiled in controversy over allegations that she had cheated in the 2004 presidential election and that her husband, son and brother-in-law are receiving kickbacks from illegal gamblers (Kirk 2005).

Quantifying corruption

The most widely respected metric for measuring corruption is the Corruption Perceptions Index developed by the Berlin-based NGO Transparency International (Diamond 1999). In the Corruption Perceptions Index, countries are ranked from a score of ten ("highly clean") down to a score of zero ("highly corrupt"). In Table 5.2 the 2010 Corruption Perceptions Index score for the Philippines is presented along with those of Australia and Canada, the home countries of many mining companies operating in the archipelago (Transparency International 2011). Corruption is substantially higher in the Philippines than in Australia and Canada. When all of the world's countries were ranked, Canada was ranked sixth out of 180 while Australia was ranked eighth out of 180; the Philippines, in contrast, was ranked 134 out of 180.

Given the high levels of corruption prevailing in the Philippines, can the regulatory environment in the archipelago even be remotely compared

Table 5.2. Comparative corruption: Philippines, Australia and Canada

Country	2010 Corruption Perceptions Index score	2010 Corruption Perceptions Index rank
Australia	8.7	8
Canada	8.9	6
Philippines	2.4	134

Source: Transparency International (2011).

to that found in Australia or Canada? An Australian or Canadian mining company encountering criticism in its home country about its activities in the archipelago will confidently assert that it has complied with all of the regulatory requirements imposed on it in the Philippines. However, given the high levels of corruption prevalent in the islands, such an assertion carries an element of disingenuity. Indeed, corruption is so pervasive in the Philippines, many view it is being an impediment to the implementation of responsible mining in the archipelago. Corruption has been "described as 'traditionally notorious' in the context of the extractive sector" (Doyle 2009, 56). "The DENR itself has been described as one of the most graft-ridden and corrupt agencies in the Philippines" (Hatcher 2010, 14). Father Romeo Catedral, the social action director of the Diocese of Marbel (where the Tampakan Project is located), opined a view that "Mining is problematic given the poor governance in the Philippines" (Catedral 2005, interview). "How can one have faith in the Philippine government to regulate mining given the corruption in the country?" (Catedral 2005, interview) In the opinion of Mayor Jerry de la Cerna, "there can be no 'sustainable' or 'responsible' mining with a corrupt government" (De la Cerna 2005, interview). Engineer Virgilio Perdigon stated, "It is cheaper to bribe monitoring officials than to try and stop acid mine drainage" (Perdigon 2009, interview).

Corruption and consent

Corruption becomes a concern of paramount importance whenever consent for mining is required from local governments units, within the aegis of the Local Government Code of 1991, or from indigenous cultural communities, under the auspices of the Indigenous Peoples Rights Act. A common Filipino euphemism for a bribe is to call it a "standard operating procedure" or "SOP." When Justina Yu was the mayor of the municipality of San Isidro in the province of Davao Oriental (from 1992 to 2001), she received three helicopter visits from mining engineers employed by foreign companies

(Yu 2005). During each of these visits, the engineers promised her "an SOP for every kilogram extracted" in exchange for the consent of the municipal government to their mining activities (Yu 2005, interview). Such anecdotal evidence of mining project proponents bribing local governments caused Father Romeo Catedral to find it disturbing that the local governments in the vicinity of the Tampakan property provided their consent "in a closed door meeting on mining company property" (Catedral 2005, interview). To Father Catedral, this was an inappropriate and opaque method of obtaining local government consent (Catedral 2005). Similar allegations also exist that Toronto Ventures Incorporated (TVI) Pacific, the Canadian operator of the Canatuan Mine, obtained the free prior and informed consent of the indigenous Subanon by bribing members of their community (International Center for Human Rights and Democratic Development 2007; Sanz 2007; Vidal 2004, 2005). Obtaining the consent of a local government unit or indigenous cultural community through corruption gravely undermines the ability of the residents of that community to participate in a meaningful manner. Instead of the consent manifesting the genuine desire of a fully informed community to accept a potentially hazardous activity knowingly, the consent simply becomes a stark display of the mining company's purchasing power. Such improperly obtained consent makes an utter mockery of the concept of free, prior and informed consent, an essential component in reducing risk, avoiding unfairness and eliminating exploitation (Shrader-Frechette 1993). Ultimately, such "bought and paid for consent" destroys the individual human autonomy of those who live in such communities. Instead of these people being able to minimize or control the hazards they are exposed to, they find themselves confronted with a hazardous activity being imposed on them that is highly capable of thrusting them from subsistence into destitution.

Low levels of civil liberties in the Philippines

The fifth factor contributing to a weak governance of mining are the low levels of civil liberties in the Philippines. To the World Resources Institute (2003), freedom of expression and civil liberties are essential components of governance when mining projects are located in hazardous areas. Civil society organizations can contribute to the entire range of disaster prevention and disaster mitigation measures (Wisner et al. 2004). "However, such citizen protest and lobbying only finds full expression under conditions of economic and political democracy and respect for human rights" (Wisner et al. 2004, 371). "Before the most vulnerable can risk becoming publically involved, there must also be a certain minimum level of social peace" (Wisner et al. 2004, 373).

Table 5.3. Comparative civil liberties: Philippines, Australia and Canada

Country	2010 Freedom House civil liberties score
Australia	1
Canada	1
Philippines	3

Source: Freedom House (2010).

Unfortunately, the Philippines is a nation with a reputation for a low level of social peace demonstrated by its low levels of political freedom and civil rights. The best available empirical indicator of liberal democracy is the freedom survey carried out annually by Freedom House, a Washington DC–based NGO (Diamond 1999). Freedom House rates countries in terms of their political freedom and their civil liberties and gives countries a score ranging from one (free) to seven (not free) for these two variables. In Table 5.3, the civil liberties scores for the Philippines is presented along with those of Australia and Canada, the home countries of many mining companies operating in the archipelago.

The Philippines has a civil liberties score of three, in contrast to Australia and Canada which both have scores of one, thus indicating a relatively lower level of civil liberties in the archipelago than in Australia and Canada (Freedom House 2010). An Australian or Canadian mining company encountering criticism in its home country about its activities in the archipelago will unhesitatingly assert that it has acquired the broad consent of the population residing where its mine is located and that it has not encountered any opposition to its project. However, given the low levels of civil liberties prevalent in the Philippines, such an assertion also carries an element of contrivance. There may well be numerous actors who object to the presence of the mine yet feel intimidated into keeping their objections to themselves. Alternatively, there may be actors offering their consent to the mine to prevent, or mitigate, harassment for having withheld their consent.[12] Attention now turns to a substantial factor impacting upon civil liberties in the Philippines: the wave of extrajudicial killings plaguing the archipelago.

---

12 An example of this comes from the municipality of Gonzaga in the province of Cagayan where the CPP alleges the Mayor attended a meeting of *barangay* captains accompanied by a platoon of armed soldiers from the Philippine Army on 8 January 2011 and on 12 January 2011. Those *barangay* captains who did not consent to mining were declared to be NPA supporters (Communist Party of the Philippines Information Bureau 2011).

The specter of these killings casts a sinister and dark shadow over all of the aforementioned aspects of weak governance.

The extrajudicial killings

The concept of extrajudicial killings

In the Philippines, Supreme Court Administrative Order No. 25–2007 defines extrajudicial killings as "killings due to the political affiliation of the victims; the method of attack; and involvement or acquiescence of state agents in the commission of the killings" (Parreno 2010, 39). Extrajudicial killings are by no means a new occurrence in the islands and they have gone on since the American colonial period (McCoy 2009). Extrajudicial killings are also something that is not confined to people involved in social activism as demonstrated by the killing of street children, petty criminals and drug dealers in Davao City (Human Rights Watch 2009). However, since the ascension to power of President Gloria Macapagal-Arroyo in 2001, there has been "an alarming spike in extrajudicial killings" (Hutchcroft 2008, 141). Most of those killed have been activists (Figure 5.3) representing groups such as peasants campaigning for land reform, labor organizers, activists against illegal logging and human rights activists (Alston 2007, 2009; Holden 2009a; McCoy 2009; Pratt 2008; Sales 2009).

Figure 5.3. Protest against extrajudicial killings, Davao City, 10 December 2008

Photo credit: Keith Bacongco.

According to Parreno (2010, 11), "most cases of extrajudicial killings involve the killing of leftist activists." Girlie Padilla, international liaison officer of Karapatan, indicated that the victims tend to be members of civil society organizations on the left of the political spectrum (Padilla 2007). Audrey Beltran, the public information officer of the Cordillera Human Rights Alliance, indicated that those killed often to belong to progressive people's organizations, such as peasant groups or labor groups (Beltran 2007). Santos Mero, the deputy secretary general of the Cordillera Peoples Alliance, made it clear that the victims tend to be activists and that "the more vocal people are, the more vulnerable they become" (Mero 2007, interview). Both men and women have been targeted and the victims have included community organizers, church workers, human rights activists, local government officials and political activists (Amnesty International 2006). "The majority of targets are people who are lawfully criticizing governmental policies by means of peaceful measures such as speeches, writing and mobilizing people" (Human Rights Now 2008, 6).

There is a widespread consensus that these killings can be attributed to the government (in general) and to the Armed Forces of the Philippines (AFP) (in particular) as opposed to just being random acts of violent crime. "Murder in the Philippines," wrote Sales (2009, 328) "is not a willy-nilly phenomenon. It is part of a carefully contrived battle plan and much can be learned by studying its shape and form." Human Rights Watch (2007, 25) similarly held the state responsible concluding that "our research, based on accounts from eyewitnesses and victims' families, found that members of the AFP were responsible for many of the recent unlawful killings." Franco and Abinales (2007, 315) concluded "agreement is widespread that the killings have AFP written all over them."

The magnitude and location of the killings

The Philippine human rights NGO Karapatan[13] is an NGO described by Pratt (2008, 774) as "highly visible and widely respected internationally." Karapatan estimates that from 21 January 2001 until 30 June 2010 the number of those killed totaled 1,206 (Karapatan 2010). While Karapatan's assessment of the number of killings is higher than those provided by others,[14] Philip Alston, the United Nations Special Rapporteur on extrajudicial executions, found

13 *Karapatan* is the Tagalog word for "right."
14 Attorney Al Parreno, in contrast to Karapatan, estimated that from 2001–10 there were only 390 extrajudicial killings in the Philippines (Parreno 2010). Nevertheless, Parreno (2010, 5) commented that the "real number of extrajudicial killings in the Philippines escapes exact determination. Regardless, however of the true body count, the mere fact that there are so many extrajudicial killings is by itself a cause for alarm."

Figure 5.4. Year by year breakdown of extrajudicial killings

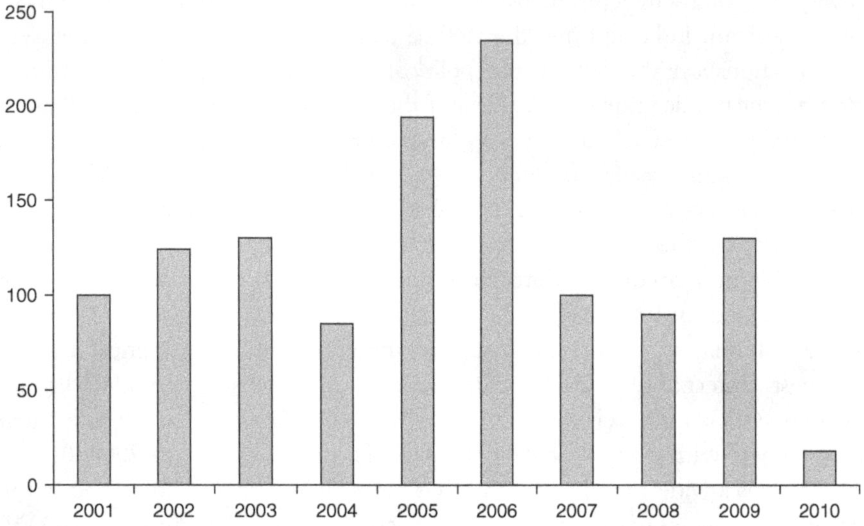

Source: Karapatan (2010).

Karapatan's estimate credible and attributed it to Karapatan's extensive geographical coverage across the archipelago (Alston 2007). A year-by-year breakdown of the killings is provided in Figure 5.4.

Although the number of killings spiked during 2005 and 2006, Alston found the number of killings to "still remain a cause for great alarm" (Alston 2009, 5). The killings have occurred in all regions of the Philippines with the Bicol Region, the Southern Tagalog Region,[15] Central Luzon and the Eastern Visayas counting for almost 55 percent of all killings (Figure 5.5).

The methodology of the killings

Most of the killings seem to follow a similar methodology wherein the victims are shot in broad daylight by men wearing ski masks riding motorcycles (Parreno 2010). After being shot, nothing is taken from the victims and they are left to die where they have been shot (Padilla 2007). The brazen nature of these attacks indicates that the assailants had little fear of any police or government

---

15 Karapatan only provides its data on a region-by-region basis not on a province-by-province basis. For the purposes of assembling its data, Karapatan combines Region IVA CALABARZON (Batangas, Cavite, Laguna, Quezon and Rizal) and REGION IVB MIMAROPA (Marinduque, Mindoro Occidental, Mindoro Oriental, Palawan and Romblon) together in one omnibus Region IV which is referred to in Figure 5.5 as the "Southern Tagalog Region."

Figure 5.5. Extrajudicial killings 2001 to 2010 (by region)

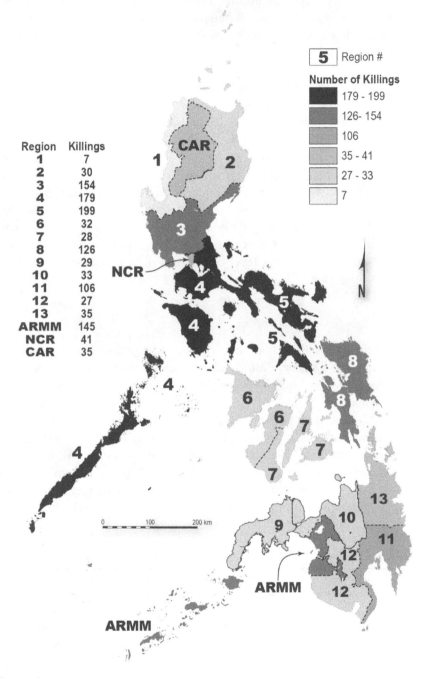

Source: Based on data from Karapatan (2010).

reaction (Human Rights Now 2008). According to the Melo Commission (2007, 5), an independent commission created by the government to address the killings, "victims were generally unarmed, alone or in small groups and were gunned down by two or more masked or hooded assailants, oftentimes riding motorcycles."

Girlie Padilla stated that before someone is killed they are subjected to target research by the AFP (Padilla 2007). According to Audrey Beltran, there is usually a 3-month surveillance period and the victims experience extensive surveillance during this time (Beltran 2007). Usually the AFP conducts this surveillance by attending rallies and photographing those who speak; Beltran has personally been photographed. Santos Mero indicated that members of the AFP wearing civilian clothes will attend rallies and will photograph speakers and they always confirm the names of those they photograph (Mero 2007). The presence of these men is obvious in that they are strangers who are never known in the communities where the rallies occur, yet they will attend numerous rallies. Mero has taken photographs of these men and the Philippine National Police (PNP) confirmed to him that they are indeed members of the AFP.

The role of Major General Jovito Palparan

A noteworthy observation about three of the leading regions for extrajudicial killings (Southern Tagalog, Eastern Visayas and Central Luzon) is that they were all areas where Major General Jovito Palparan, an AFP officer, was assigned (Table 5.4).

From May 2001 until April 2003, (then) Colonel Palparan was the commander of the 204th Infantry Brigade on the island of Mindoro (Melo Commission 2007). While Palparan was in command on Mindoro, he implemented operational plan "Habol Tamaraw" (Buffalo Hunt) and this led to so many killings of activists that Jun Saturay, a former member of the Alliance Against Mining in Oriental Mindoro, lost count of how many

Table 5.4. The postings of Major General Jovito Palparan

Time Period	Military Unit	Region
May 2001 to April 2003	204th Infantry Brigade	Southern Tagalog
February 2005 to August 2005	8th Infantry Division	Eastern Visayas
September 2005 to September 2006	7th Infantry Division	Central Luzon

Source: Melo Commission (2007).

people had been killed (Saturay 2007). After his promotion from colonel to major general, Palparan was assigned to the Eastern Visayas (the islands of Leyte and Samar) and given command of the 8th Infantry Division from February 2005 until August 2005 (Melo Commission 2007). According to Maria Patalinghug-Vasquez, Central and Eastern Visayas coordinator of Task Force Detainees of the Philippines, this generated a substantial amount of trepidation among many members of civil society. During this time, human rights activists were often afraid to go to the Eastern Visayas on fact-finding missions (Patalinghug-Vasquez 2005). In September 2005, Palparan was transferred from the Eastern Visayas to Central Luzon and given command of the 7th Infantry Division, a posting held until his retirement from the AFP in September 2006 (Melo Commission 2007). Wherever Palparan has been assigned, successive killings have taken place and there have been cases where witnesses have identified soldiers under his command as being the perpetrators of the killings (McCoy 2009). Even though Palparan has never admitted responsibility for any killings where he has been in command, he has made it clear that he may have inspired them (Melo Commission 2007).

The killings of antimining activists

Many of the victims of extrajudicial killings have been people engaged in activism against large-scale mining. One of the earliest killings of antimining activists occurred on 8 April 2002 when Expedito Albarillo and his wife Manuela, two antimining activists in San Teodoro, Oriental Mindoro, were forcibly removed from their house by a group of eight armed men and, despite their pleas for mercy, killed in front of their 8-year-old daughter (Human Rights Now 2008). This killing occurred while then colonel Palparan was commanding the 204th Infantry Brigade on Mindoro. On 5 February 2005 Abe Sungit, an antimining activist in Roxas, Palawan, was shot and killed by two unidentified gunmen riding on a motorcycle (National Democratic Front of the Philippines 2007). Two weeks later, Joel Pelayo and Rodel Abraham, antimining members of the Central Luzon Aeta Association, were shot and killed while visiting Balanga, Bataan (Kalikasan 2006). Eighteen days after that, Romy Sanchez, a member of the Save the Abra River Movement was killed in Baguio City; Sanchez had been campaigning against large-scale mining being conducted in the CAR by Lepanto Consolidated Mining (Kalikasan 2006). On 20 August 2005 Reverend Raul Domingo, a minister in the United Church of Christ and an ardent environmentalist who had campaigned against the Palawan Nickel Project (Figure 2.2 and Table 2.1), was killed in Puerto Princesa City, Palawan (Kalikasan 2006).

The year 2006 saw a spike in the overall numbers of extrajudicial killings (Figure 5.4) and antimining activists were by no means immure to this. On 27 May 2006 Noli Capulong, an outspoken opponent of mining in the Southern Tagalog Region, was shot and killed by motorcycle riding gunmen in Calamba City, Laguna; Capulong had been the spokesperson for the Southern Tagalog Environmental Action Movement and had denounced projects detrimental to the environment (IBON 2006b). Five days later Rogelio Lagaro, a member of the Labugal Tribal Association, was killed in Colombio, Sultan Kudarat; Lagaro had been an opponent of the Tampakan Project (Kalikasan 2006). One week later, Markus Bangit, coordinator of the CPA Elder's Desk and an active opponent of large-scale mining, was shot at a rest stop in San Isidro, Isabela, while travelling by bus from Tabuk, Kalinga to Baguio City; Bangit was shot five times by men wearing ski masks who escaped in a van which had been following the bus (IBON 2006b). Twelve days after the death of Bangit, Eladio Dasi-an, vice chair of Defend Our Patrimony-Negros, was gunned down in Guihulngan, Negros Oriental (Kalikasan 2006). On 31 July 2006, Rei Mon Guran from Aquinas University of Legazpi, was gunned down by two motorcycle riding assailants while boarding a bus; Guran had participated in rallies against the Rapu-Rapu Polymetallic Project during June and July of 2006 (Oxfam Australia 2008). Three days later, Pastor Isias de Leon Santa Rosa, a writer of articles criticizing the Rapu-Rapu Polymetallic Project, was killed by armed men outside of his house in Daraga, Albay (Human Rights Watch 2007). On 12 December 2006, Attorney Gil Gujol was gunned down in Sorsogon; Gujol was representing a group of complainants from Sorsogon who had filed a class action lawsuit demanding compensation for damages caused by the October 2005 tailings spills (Oxfam Australia 2008). Later that same month, Manuel Balani was stopped and killed by a group of men in Albay; before being shot Balani was asked, "So you are the one who does not want the mine to open?" (Human Rights Watch 2007, 40).

During 2007 and 2008 the killings continued, and on 3 October 2007 Armin Marin was gunned down while leading an antimining rally in front of the office of the Romblon Nickel Project on Sibuyan Island in the province of Romblon (Figure 2.2 and Table 2.1); Marin's death was witnessed by hundreds of people (Kalikasan 2011). On 2 December 2007 Dodong Ledesma, a prominent opponent of large-scale mining was shot and killed in San Miguel, Surigao del Sur (Legal Rights Center 2007). On 23 December 2008 Fernando "Dodong" Sarmiento, secretary general of Panalipdan, New Bataan, was shot five times and died in Compostela Valley; Sarmiento had been involved in activism against the Tagpura Copper Project (Figure 2.2 and Table 2.1) and had been interrogated by the AFP during July 2008 about his involvement in antimining activism (Pinoy Press 2008).

A particularly disturbing killing occurred when, on 9 March 2009, Eliezer "Boy" Billanes, an antimining activist with the nongovernmental organization SOCCSKSARGENDS-AGENDA (South Cotabato, Cotabato, Sultan Kudarat, Sarangani, General Santos City and Davao del Sur Alliance for Genuine Development), was shot and killed by two unidentified men riding a motorcycle in Koronadal City, South Cotabato (MindaNews 2009).[16] Billanes was involved in opposing "large-scale, extractive, commercialized and aggressive mining" (Billanes 2005, interview). Three years earlier, drawing attention to deforestation in the province of South Cotabato (Figure 4.12), Billanes warned of the vulnerability of the vicinity of the Tampakan Project to landslides and flooding stating, "We are in a situation where we are waiting for [a] disaster" (MindaNews 2006). On 22 April 2009 Ludinio Monson, an antimining activist in Davao Oriental, was shot and killed in Boston, Davao Oriental (GMA News 2009).

The end of the twenty-first century's first decade saw no abatement of the killings and, on 10 February 2010 Ricardo Ganad, a staunch opponent of the Mindoro nickel project (Figure 2.2 and Table 2.1), was shot and killed by two men while he was standing on the terrace of his house in Victoria, Oriental Mindoro (GMA News 2010). Nineteen days later two unidentified gunmen shot and killed antimining activist Gensun Agustin in Calamegatan, Cagayan (United States Department of State 2011). On 16 May 2010 Mike Rivera, a former provincial administrator, was shot and killed by two unidentified gunmen while entering church in Calapan City, Oriental Mindoro; Rivera had been active in campaigning against the Mindoro nickel project (Virola 2010).

The end of the Macapagal-Arroyo presidency did not necessarily bring an end to the killings either, and on 24 January 2011 Dr Gerry Ortega, a prominent environmentalist, radio broadcaster and veterinarian, was shot and killed in Puerto Princesa City, Palawan. Ortega, a former director of the Crocodile Farm, was an outspoken opponent of large-scale mining (Silverio 2011). On 13 April 2011 Santos Manrique, chair of the environmental group Panalipdan and an opponent of the King-King project (Figure 2.2 and Table 2.1), was killed at his house in Pantukan, Compostela Valley, by an unknown gunman; earlier Manrique declined to participate in the scoping session for the mine's environmental assessment because he felt that the situation was becoming too dangerous for him (Lacorte 2011a).

Many antimining activists have also received death threats regarding their activism. Prior to Eliezer Billanes' death, he had received several death threats and was believed to be the object of surveillance by the AFP; indeed, on the day

16 The authors found the killing of Billanes particularly disturbing since the senior author had interviewed Billanes in Koronadal City, South Cotabato on 18 May 2005.

of his death he had just attended a dialogue with the AFP regarding his safety (MindaNews 2009). When Jun Saturay was involved in antimining activism in Oriental Mindoro, his father had received information that "in order to stop the antimining campaign 'one would be sacrificed'" (Saturay 2007, interview). Eventually after receiving numerous threats and after seeing several other antimining activists killed, Saturay sought refugee status in the Netherlands (Saturay 2007). Camilo Guisadio, social action director of Bula Parish in General Santos City, stated that during the antimining campaign carried out by the Diocese of Marbel, he had been listed as number two on the order of battle of the AFP for the province of South Cotabato – meaning that he was second on the list to be assassinated by the AFP in South Cotabato (Guisadio 2007).

Some antimining activists have also been subjected to black propaganda designed to intimidate them into a cessation of their activism. Belén Gallego, a member of the environmental group Panalipdan, to which Santos Manrique also belonged, began to hear rumors in Pantukan, Compostela Valley that she had been killed or abducted (Lacorte 2011b). When a fact-finding mission was sent to Pantukan by the church group Exodus for Justice and Peace on 21 May 2011, it was stopped by a joint AFP/PNP roadblock despite the fact that the mayor of Pantukan had provided it with a written permit authorizing its fact-finding mission (Lacorte 2011b).

To Rovik Obanil, the extrajudicial killings have imposed a chilling effect upon the advocacy of the Legal Rights and Natural Resources Center; some members of their partner organizations have been killed and harassed and one cannot help but think that it disrupts their work (Obanil 2009). Jesus Garganera has no doubt that the extrajudicial killings are having a chilling effect on civil society; there have been several ATM members who have been the victims of extrajudicial killings and several ATM members have found their names on the order of battle of the AFP (Garganera 2009). Amalie Obusan related how the partner organizations of Greenpeace Southeast Asia in South Cotabato feel threatened; Greenpeace Southeast Asia visited South Cotabato just after the killing of Billanes and there was a detectable reluctance to speak up on antidevelopment issues (Obusan 2009). The killings of antimining activists are definitely an example of weak mining governance in the Philippines. If speaking up against mining projects may result in fatal consequences, one will certainly give such activism a second thought. This further limits the ability of communities living near mines to impact their development.

The extrajudicial killings as a violent dimension of neoliberalism

In Chapter Two the role of the neoliberal economic development paradigm was discussed as an impetus behind the government's aggressive promotion of mining. Worldwide neoliberalism has generated numerous protest movements that have often been "ruthlessly put down by state powers" (Harvey 2003, 189). States often engage in systematic repression to "ruthlessly check activist movements" that challenge neoliberal policies (Harvey 2005, 200). The killings of antimining activists in the Philippines are an example of such ruthless repression of neoliberalism's opponents; as Rodriguez (2010, 15) wrote, "Total war [has become] necessary for the Philippines to continue to implement neoliberal economic policies and to coerce the Filipino people into accepting them."

Many of the victims of extrajudicial killings were activists against projects and policies implemented in adherence to the government's neoliberal development paradigm. According to Audrey Beltran, the victims are often critical of development projects (such as agribusiness plantations, export processing zones and mines) and development policies (such as bilateral trade agreements); the killings are an attempt to silence criticism of these projects and policies by killing their opponents (Beltran 2007). Girlie Padilla stated that the AFP is often deployed in areas where multinational corporations have projects; this is done to eliminate opposition to these projects as "less opposition from below means less opposition to the project being implemented" (Padilla 2007, interview). According to Jun Saturay, a large percentage of extrajudicial killings occur in areas where development projects are located; many of the victims are activists from community organizations opposing these projects and the government has reverted to a policy of "elimination by assassination" (Saturay 2007, interview). Daniel Conejar, Mindanao coordinator of Task Force Detainees of the Philippines, indicated that the opposition of people to bilateral agreements, such as the Japan-Philippines Economic Partnership Agreement, is silenced by having them executed (Conejar 2007). To Carlos Conde, the advocacies of the victims is a glaring example of how neoliberalism creates violence; whenever activists are killed, one will find a powerful neoliberal interest being opposed (Conde, 2009).

Multinational corporations deny having anything to do with the killings of activists who oppose their projects but, nevertheless, they still gain from having the AFP do it for them. Father Frank Nally is a priest from the Society of St Columban (a religious order with a long history of activism in the archipelago) who spent nine years on the island of Mindanao. Father Frank also views these killings as a policy of "elimination by assassination" stating, "Mining companies are making a killing because of the killings" (Nally 2007, interview). Kelly Delgado, the Karapatan representative for Southern Mindanao, pointed out how there have been instances where union representatives at banana plantations have been killed because the government is trying to eliminate unions. If the multinational corporations operating agribusiness plantations are not behind these killings, they at least benefit from them by receiving a union free workforce (Delgado, 2007). In the export processing zones of Luzon and Cebu, union organizers are frequently killed. As Legaspi (2007, 107) wrote, "Union formation has become an activity as difficult as a carabao passing through the eye of a needle."

Since the opposition faced by mining projects is eliminated though violence, neoliberalism – a utopian project envisaging the achievement of economic order through freedom – actually constitutes an aggressive form of development denying freedom to those who stand in its way. Proponents of neoliberalism view it as a process where providing global capital unimpeded access to a developing country's economy will generate prosperity and, ultimately, an improvement in the welfare of that nation's citizenry. In the Philippine context, however, providing global capital unimpeded access to the archipelago's economy has generated controversy that is squelched by killing those who object to it. This hardly resembles a situation consistent with the lofty principles of freedom and liberty upon which neoliberalism is ostensibly predicated.

From Technological Solutions to Risk Society

The government of the Philippines and the mining industry are by no means oblivious to the risks presented by the intersection of mining with natural hazards and these risks are a serious cause for concern to them. Their solutions to these risks lie in the use of two highly technocratic measures: subjecting mining projects to an environmental impact assessment (EIA) and the use of state of the art technology. Both the government and the mining industry maintain that the EIA system will identify the risks posed by natural hazards and that multinational mining companies have the skills, experience and technology to minimize these risks to an inconsequential level and consequently nothing can go wrong. Consider, for example, this confident description of the skills and abilities of the Marcopper Mining Corporation made by Lopez (1992, 375): "Marcopper has been recognized as a leader in the area of corporate responsibility, particularly in the fields of environmental protection, safety and community relations."[17] Four years after Lopez wrote those hubristic words, the worst environmental incident ever sustained in the Philippines took place at the very mine being written about.

Most people in the Philippines involved in antimining activism, however, lack faith in such technocratic solutions to the problems posed by the intersection

---

17 Lopez's book, *Isles of Gold: A History of Mining in the Philippines* (1992) was commissioned by the Chamber of Mines of the Philippines.

of mining, natural hazards and vulnerable populations. When Filipinos hear mining industry insiders making statements echoing the bravado of Lopez, they are immediately reminded of what transpired on the island of Marinduque and such technocratic assurances ring hollow. The weak EIA system imposed upon mining projects (with its inadequate facilities for public participation), and the limited governance of mining in the archipelago (with the high levels of corruption and weak civil liberties) make it extremely difficult for these people to have any meaningful input into how mining projects are carried out. Indeed, for many of these people, activism against large-scale mining ultimately proved fatal. Having discussed the inadequacies of technocratic solutions, attention now turns to how the government's ambitious implementation of a mining-based development paradigm constitutes an example of risk society. In risk society, harm comes not from what the forces of nature may do to humans but from what humans may do to themselves with their complicated technology. As the technologies used by humans have become increasingly more complex, humans have transformed themselves from being at the mercy of nature to being at the mercy of their complicated technology; something they may be powerless to address if and when it malfunctions.

Chapter Six

RISK SOCIETY IN THE PHILIPPINES

Technocratic Solutions as Modernity

The views expressed in the preceding chapter by the Philippine government and the mining industry that technology can prevent any and all disasters is "essentially a western materialistic interpretation of nature, where disaster is seen as a disruption to normal life" (Bankoff 2003a, 177). According to this perspective, disasters are the result of the arbitrary and capricious forces of nature and the logical response to such threats lies with the scientific prediction of their occurrence and the implementation of technocratic measures to reduce their risks. "Disaster prevention, therefore, is seen as largely a matter of improving scientific prediction, engineering preparedness and the administrative management of hazards" (Bankoff 2001, 24). Much like the complete confidence shown by the government in its adherence to neoliberal economic theory, this articulation of complete confidence in humanly engineered solutions to the problems posed by natural hazards exemplifies modernity (Holden and Norman 2009; Holden 2011).

An essential component of modernity is trust and confidence in experts; "modern life [is] dominated by knowledge and science" (Harvey 1990, 15). "The nature of modern institutions is deeply bound up with the mechanisms of trust in abstract systems, especially trust in expert systems" (Giddens 1990, 83). Under conditions of modernity, the world is bifurcated into two distinct groups: experts and nonexperts (Holden and Norman 2009; Holden 2011). The elite of the experts are scientists; they constitute an expert vanguard offering the stark alternatives of salvation through them or chaos. Under conditions of modernity, only scientists can accurately know what risks really are and how these may be overcome; in the eyes of the technological elite, the public are ignorant and clueless (Beck 1992). Scientists "determine risks" while the population "perceive risks;" any deviations from this pattern indicate the extent of "irrationality" and "hostility to technology" prevalent among the uneducated masses (Beck 1992).

Mining as Ecological Modernization

With technology, environmental problems will not be obstacles to economic growth

Ecological modernization is a view that, with appropriate use of technology, environmental problems will not be obstacles to economic growth. This paradigm "assures us that no tough choices need to be made between economic growth and environmental protection" (Blowers 2003, 65). "Ecological modernization theorists acknowledge the material/ecological crisis that has emerged from contemporary capitalism/industrial institutions, yet maintain the crisis can be resolved through appropriate technology and industrial retooling without significantly changing the nature of existing political economic institutions" (Emel and Krueger 2003, 10).

Mining: A discourse of ecological modernization par excellence

The mining industry, in particular, has become "a vehicle for the more general argument that environmental risk can be managed, that society should therefore not be afraid in the face of such risk and that public risks are best managed privately" (Bebbington et al. 2008, 900). The mining industry likes to depict itself as having undergone a profound transformation from the "old mining" to the "new mining" (Bebbington et al. 2008). The old mining damaged the environment, had dangerous workplaces and operated in complete disregard for the needs of communities adjacent to its projects. The new mining is socially and environmentally responsible and uses technology to ensure the management (but not complete elimination) of environmental risk. To Bebbington et al. (2008, 900) the claims of the mining industry that its use of technology can prevent any, or all, disasters "constitute a discourse of ecological modernization *par excellence.*"

Just as under conditions of modernity, in general, scientists reject the views of members of the public as irrational and ill informed. Under conditions of ecological modernization, in particular, the mining industry rejects the views of those lacking confidence in its ability to minimize hazards through technology (Holden 2011). The mining industry, in its opinion, is staffed by experts who understand the risks they are encountering; it has the technology, skills and abilities to reduce these risks to an inconsequential level. Those lacking confidence in the ability of the mining industry to minimize risks are simply making it clear that they lack knowledge of the industry's abilities. Otherwise, say mining industry leaders, these people would cease objecting to the presence of mining operations near their communities and would welcome these operations with open arms as harbingers of prosperity. According to Whitmore (2006, 310), the mining industry views those opposed to mining

as consisting of "ignorant and 'antidevelopment' communities and NGOs."
An example of such a view comes from Chris Hinde, editorial director of
Mining Journal, who wrote that when a mine is proposed its "potential physical
impacts appear to be measurable, predictable and manageable to mining
industry insiders" (Hinde 2006, 2). However, when the same mine is proposed
to people who are not mining industry insiders, "there are powerful social
factors that all too frequently lead to an inability to recognize the real nature
of the risks that come with mineral extraction" (Hinde 2006, 2). To Hinde,
these people lack expertise in the mining industry's skillful use of technology
and they will "tend to see [mining] as a threat to personal health and safety"
and will "say 'no' to [mining]" (Hinde 2006, 2).

Risk Society: A Rejection of Modernity

Traditional society, industrial society and risk society

The view expressed by the mining industry that technology can resolve all
problems – and the concomitant rejection of this view cited in Chapter Five
by those opposed to mining – demonstrates an idea that has begun to occupy
center stage within debates over ecological modernization: the concept of risk
society (Holden 2011). According to Beck (1992), humans have gone through
three phases of social development: traditional society, industrial society and
risk society.

Traditional society: Risks beyond the ambit of human control

In traditional society, risks take the form of natural hazards that are unavoidable
and are invariably assigned to the agency of supernatural forces. Hazards
are things such as storms, floods, famines, droughts and plagues – all things
beyond the ambit of human control. In traditional society, "nature and the
unbroken compulsions of tradition are to blame for the sicknesses, crises and
catastrophes from which people suffer" (Beck 1992, 159). The classic example
of risks inherent in traditional society would be the Black Death of 1348,
an event (given the quality of fourteenth-century epidemiology) completely
beyond the comprehension of humans living at that time.

Industrial society: Risks emanating from wide-scale social forces

Traditional society is followed by industrial society wherein the characteristics
of risk change to become contingent upon the actions of wide-scale social
forces and the predominant concern is the creation and distribution of wealth
(Beck 1992). Hazards in industrial society are things such as mass unemployment
resulting from the failure of the economy. The archetypal example of the risks

inherent in industrial society was the Great Depression of the 1930s when the social order almost collapsed due to the severe disruption of the global economy.

Risk society: The inability of humans to control their own technologies

Industrial society was, in turn, followed by risk society where risks emanate from the inability of humans to control their own technologies. Beck (1992, 21) defines "risk," in such conditions, as "a systematic way of dealing with hazards and insecurities induced and introduced by modernization itself." As opposed to "externally caused dangers (from the gods or nature), the historically novel quality of today's risks derives from internal decision" (Beck 1992, 155). "In contrast to all earlier epochs,' wrote Beck (1992, 183), "the risk society is characterized essentially by a lack: the impossibility of an external attribution of hazards." In the risk society, "new technologies" have placed humanity "balancing on the verge of catastrophe" (1992, 185). The most notorious example of the risks inherent in risk society would be the 1986 nuclear accident at Chernobyl, in Ukraine, which caused radiation to spread as far as Ireland. This accident demonstrated how a technology developed by humans can become as great, or greater, a threat to humans than the forces of nature can ever be. Perhaps the quintessential example of risk society is anthropogenic climate change; as greenhouse gas emissions continue unabated, the earth's climate warms and this produces a whole set of hazards far more serious than humans would have encountered prior to climate change.

Large-scale mining in the Philippines: An example of risk society

The location of large-scale mining projects in the Philippines, a country beset by natural hazards, is a clear example of a risk society. The archipelago has always encountered the hazards posed by typhoons, earthquakes, tsunamis, volcanoes and El Niño–induced drought, but now the interaction of hardrock mining with these hazards has the potential to exacerbate greatly the scope and scale of the risks they pose.

An extension of risk society into new terrains

There are writers who view Beck's discussion of risk society as something confined to the developed world (Wisner et al. 2004). "Beck's work and the discussion it has stimulated are important," wrote Wisner et al. (2004, 18); however, it "is rather remote from the dynamics of hazard vulnerability and risk in [less-developed countries]." There are two reactions to this argument. First, it must be borne in mind that most of the large-scale mining projects being constructed in the Philippines are being developed – at least in part – by mining companies from the developed world and, as made clear by writers such as Warhurst (1992, 1999), these mining companies use the same technology in developing countries that they use in their home countries; the insertion of such developed world technology into the Philippines arguably constitutes an extension of risk society into new terrains. Second, Beck (1992) also makes the point in his discussion of the transformation from industrial society to risk society that the two overlap and that risks are often distributed in a stratified, or class-specific, manner with the affluent being less exposed to risk while the poor and marginalized are more exposed to risk. "Like wealth, risks adhere to the class pattern, only inversely: wealth accumulates at the top, risks at the bottom" (1992, 35).

The stratified distribution of the risks

How the poor bear the costs

In the Philippine context, examples of the people who bear the risks of mining projects are subsistence farmers, subsistence fisher-folk and indigenous peoples. Consider, for example, the residents of Barangay Macambol in the municipality of Mati in the province of Davao Oriental (Figure 6.1).

Figure 6.1. Residents of Barangay Macambol

Photo credit: The authors, published originally in the *International Journal of Science in Society*.

The residents of Barangay Macambol are subsistence fisher-folk who fish in the waters of Pujada Bay; they are quite concerned about the possible environmental effects of the proposed Pujada Nickel Project (Figure 2.2 and Table 2.1) and the implications it may have upon their livelihood. Barangay Macambol is not an area vulnerable to typhoons (Figure 4.1) but it is an area highly vulnerable to earthquakes (Figure 4.2). For example, in 1924 there was an earthquake of magnitude 8.3 just outside of Pujada Bay (Cabanlit 2007). If an earthquake causes a tailings dam to fail and contaminates the waters of Pujada Bay, the residents of Barangay Macambol will lose their livelihoods and be thrust from subsistence into destitution. Barangay Macambol is also an area vulnerable to El Niño–induced drought (Figure 4.11). Should there be water table drawdown due to mine-pit dewatering during an El Niño–induced drought, the numerous springs which drain into Pujada Bay will draw sea water into the aquifer and this will lead to a salinization of the groundwater, thus depriving the residents of Barangay Macambol of their freshwater supply (Rodolfo and Siringan 2006).

How the rich receive the benefits

In contrast to the poor and marginalized, the rich and powerful members of the archipelago's society can, in the words of Beck (1992, 35), "purchase safety and freedom from risk." Indeed, an argument can be made that the rich and powerful can actually benefit from a further marginalization and impoverishment of the lowest levels of society. Once displaced from their subsistence farming and fishing, the poor will look for work in the large cities of the archipelago or in its export processing zones. When this occurs, the entry of these people into the labor market will push wages down and benefit the rich and powerful by providing them with lower cost labor. Carlos Conde related how well over half of his recent interviews in Metro Manila were of recent migrants from the islands of Samar and Leyte (Conde 2009). In Tondo, the sprawling shantytown northwest of Manila, there is a large Waray-speaking population consisting of migrants from the Eastern Visayas who have moved to Manila looking for better opportunities. Poor landless people are forced into becoming rural to urban migrants; they flood into the export processing zones where they drive down wages. To Conde (2009, interview), these export processing zones are "enclaves of sorrow."

Similarly, just as the rich and powerful members of the archipelago can benefit from the impoverishment of the poor and marginalized, so can the rich and powerful people and the rich and powerful countries of the world. If displaced members of the rural poor migrate to export processing zones and drive down wages, they will provide more locations for multinational corporations to outsource their operations and this puts downward pressure on wages within the labor markets of the developed capitalist countries. In Chapter One the phenomenon of overseas Filipino workers (OFWs), who work abroad and then remit funds home to their families, was discussed. If subsistence fisher-folk, such as the residents of Barangay Macambol become displaced by a mining related environmental disruption, they will quite likely migrate to Manila and, once there, become OFWs. As Tyner (2009, 112) wrote:

> Manila is the destination of thousands of displaced peasants from the rural hinterland. These internal migrants either may arrive in Manila specifically for overseas employment, or perhaps will later take advantage of this opportunity once in the city. Either way, Manila-based labor recruiters benefit from a continual supply of reserve labor.

Once displaced, poor people from rural areas enter what Rodriguez (2010, 53) called the "reserve army of Philippine labor at the ready for deployment around the world." These people can then be made available to work abroad as bartenders, construction workers, maids, nannies, performers, seamen, waiters and waitresses. This will provide low cost unskilled labor to benefit developed capitalist countries and, by alleviating demands for higher wages from unskilled workers within those countries, also benefit the rich and powerful of those countries.

Civil Society Opposition to Mining: A Lack of Faith in Technology

Modernity: A concept with shallow roots in the Philippines

The Western project of modernity has not yet become fully accepted in the Philippines; "the culture of modernity and secularization that has dominated; the West has not made any deep impact on the Philippine culture so far" (Picardal 1995, 44). Across the archipelago, "people place little faith in the security of science and technology to control the forces of nature" (Bankoff 2003a, 178). This can be attributed to the long years of oppressive domination by the Roman Catholic Church (*La Frailocracia*) that occurred under the Spanish colonial regime. At this time, the church "taught a mystified and otherworldly version of Christianity to indoctrinate and subdue the masses for their conquerors" (Nadeau 2008, 30). Filipinos were made to believe that all bad things that happened to them in this world (including natural disasters) were the wrath of God and the lingering effects of this have precluded them from gaining an appreciation for science and have instilled in them an indifference to technology (Roque and Garcia 1993). This lack of trust and confidence in science and technology to solve the problems posed by locating large-scale mining projects amid the natural hazards prevalent in the archipelago underlies the opposition of civil society to the government's efforts to establish a mining-based development paradigm (Holden 2011).

Civil society in the Philippines

With the advent of globalization and the hegemonic dominance of the market-based paradigm of neoliberalism, the term "civil society" has become increasingly popular – particularly in discussions of sustainability and social change in the developing world (Holden 2010). "Civil society" can be defined as "the sector of society existing separate and apart from the state while lying between the state and the individual or family" (Holden 2010, 414). The Philippines has a thriving civil society that has one of the most active NGO movements in the world (Holden 2005b).

Social movements as vehicles for change

Writers on risk society place an emphasis on social movements as a vehicle for change. "Social movements provide significant guidelines [for] potential future transformations" (Giddens 1990, 158). Beck's views are that "Social movements raise questions that are not answered by the risk technicians" (1992, 30). Across the archipelago, social movements have emerged opposing the government's promotion of mining. According to Vivoda (2008, 134) there is "a growing constituency against large-scale mining in the Philippines." "The socio-environmental legacy of mining," wrote Hatcher (2010, 2), "has provoked the uproar of one of the most numerous and organized civil society [sectors] in the world." In the Philippines there are "strong antimining coalitions bridging the concerns of a number of actors such as indigenous communities, the Catholic Church, environmental groups as well as a wide range of other regional and national civil society organizations" (Hatcher 2010, 17). This opposition has involved protests (Figure 6.2), litigation, administrative proceedings and the implementation of antimining moratoriums by local governments banning mining within their jurisdictions (Holden 2005b).

There has also been violent opposition from the New People's Army (NPA), which has attacked mining projects, destroying equipment and killing security guards (Holden and Jacobson 2007b). This opposition has not gone unnoticed by the mining industry. One unnamed exploration company president was quoted by the Fraser Institute (2008, 24) stating: "[In the Philippines], local interest groups stop mining with backing from NGOs supported by European

Figure 6.2. Antimining protest, General Santos City, May 2005

Greenies." An unnamed mining company president was also quoted by the Fraser Institute (2011, 49) stating: "[In the Philippines], NGOs, peasants and church groups override [the] government constantly. You can spend millions developing a property in the Philippines, only to have it swept away by peasants, lobby groups [and] churches."

The opposition of the Roman Catholic Church to mining

One of the most significant and well-organized sources of opposition to mining in the islands is the Roman Catholic Church (Holden and Jacobson 2007a, 2011). For much of Philippine history, the church acted as an alienating institution doing little more than exacting tribute and providing rituals in return. However since the Second Vatican Council of 1962–65, (and particularly since its role as the preeminent bulwark against the machinations of the conjugal Marcos dictatorship) the church has developed a strong social consciousness and has shown no shyness or reluctance to advocate on behalf of the archipelago's poor and marginalized (Holden 2009b; Holden and Jacobson 2007a, 2011; Holden and Nadeau 2010; Nadeau 2002, 2005, 2008). The church is a highly influential institution in this heavily Catholic country where 81 percent of the population is Roman Catholic (Central Intelligence Agency 2011). Environmental Science for Social Change (1999, 95) described the church as "a vitally important part of the life and history of the Philippine nation; it is, in a sense, the soul of the nation; more than any other it has shaped the ethos of the nation." Arthur Neame, the East Asia program manager for the British NGO Christian Aid, described the Roman Catholic Church as being, "the foremost social institution in the Philippines" (Neame 2005, interview). It must be emphasized that this ecclesial opposition to the government's mining-based development paradigm is not confined to select locations in the Philippines or to select members of the church. The Catholic Bishops Conference of the Philippines (CBCP) has declared its opposition to mining in pastoral letters in 1998, 2006 and 2008 (Holden and Jacobson 2011). This is noteworthy as the CBCP sits at the apex of the Catholic hierarchy in the Philippines and is subservient only to the Vatican. Pastoral letters from the CBCP are of substantial significance and are not to be taken lightly since they contain guidelines and official positions extracted from the universal dogmas of the church and the encyclicals of the Pontiff; they have undergone substantial scrutiny and analyze the church's relationship with the state (Holden and Jacobson 2007a). One salient action undertaken by the church was the September 2009 statement of the Justice, Peace and Integrity of Creation Commission of the Association of Major Religious Superiors of the Philippines (JPICC AMRSP). In this statement the authors specifically cited the disaster vulnerability of the Philippines as one of the most serious problems associated with large-scale mining. As the JPICC AMRSP (2009, 2) wrote:

> Being vulnerable to earthquakes, typhoons, and droughts, the Philippines are considered by many to be one of the most disaster prone countries in the world. The propensity of these events further exacerbates the potential environmental effects of metallic and gold mining projects; it is an extant contingency to have a mine site requiring perpetual care and attention, it is a substantially more complicated contingency to have a mine site requiring perpetual care in a situation where a torrential rain storm, from a typhoon, causes a tailings dam to overflow or where an earthquake causes a catastrophic tailings impoundment failure. Couple the intrinsic geographical vulnerability of the Philippines to natural disasters with the economic vulnerability of many Filipino communities and one begins to gain a further dimension of the strong opposition to metallic and gold mining. Moreover many communities in

the archipelago are communities of subsistence farmers or subsistence fisherfolk who live a very precarious existence at the best of times.

Many members of the church are also unassuaged by the assurances of the mining industry that its skills and technologies can ameliorate the risks posed to mining by natural hazards. Consider Sister Susan Bolanio of the Oblates of Notre Dame. From 1992–95, and again from 2000–04, Sister Susan was the social action director of the Diocese of Marbel on the island of Mindanao. During this time, Sister Susan was involved in activism against the Tampakan Project. One thing the project proponent told Sister Susan (in an effort to assuage her concerns regarding mining's environmental effects) was that they have "world class technology" and that their operations would have no effect upon the environment (Bolanio 2005, interview). To Sister Susan, the view that technology can prevent environmental harm is unacceptable; humans have frailties and technology cannot solve all problems (Bolanio 2005). Also, technology certainly cannot overcome unpredictable events such as earthquakes or tsunamis; in Sister Susan's words, "No world class technology could ever replace the work of God" (Bolanio 2005, interview).[1]

Armed opposition of the New People's Army

There has also been violent opposition to mining from the NPA which, as shown in Table 6.1, has attacked mining projects, destroying their equipment and on occasion, killing mining supporters and security personnel. In January of 2011, the National Democratic Front of the Philippines (NDFP) – an umbrella organization representing the CPP, NPA and several other left-wing groups – drew attention to the interaction of mining with the natural hazards present in the islands. According to Jorge "Ka Oris" Madlos, a prominent NPA cadre on the island of Mindanao, "Banning, disabling and dismantling these large, environmentally-destructive operations are still the best ways we could avert the disastrous impact on people's lives and livelihood of calamities such as this. They are legitimate military targets of NPA attacks" (Ellorin 2011). Madlos also added that stopping these operations through military action by the NPA may be a form of disaster prevention. "Prevention is still better than curing the aftermath damages of these calamities" (Ellorin 2011).

---

1 In this regard, Sister Susan's views echo those of the Brazilian liberation theologian Leonardo Boff, a spiritual and intellectual influence on her (Bolanio 2005). Boff (1997, 65) rejected technological solutions to environmental problems asking, "Is it not an illusion to think that the virus attacking us can be the principle by which we will be made well?"

Table 6.1. Attacks on mining projects carried out by the New People's Army

Date	Details of attack
4 July 2000	Members of the NPA attacked the facilities of the Mindoro Nickel Project operated by the Crew Development Corporation, killing a local supporter of the project as well as eight members of the PNP who were sent to investigate. Buildings were set on fire and equipment was stolen (Eraker 2000).
3 October 2007	Forty NPA cadres attacked the Labo Gold Project operated by Australia's El Dore Mining. Heavy equipment and vehicles were burned and the mine's security guards had their weapons taken from them. The NPA stated that the raid was conducted to punish the mining industry for damaging the environment, displacing villagers and exploiting resources which should benefit Filipinos (Gomez 2007).
1 January 2008	Members of the NPA attacked the Tampakan Project to punish the mining company for engaging in land grabbing, plundering and environmental destruction. The attack was also declared to be an important milestone in the effort to defend the ancestral domain of the B'laan tribe (Holden et al. 2011).
21 December 2008	Approximately sixty NPA cadres raided the La Fraternidad Project operated by San Roque Metals. Security guards were disarmed and three backhoes were set on fire (Caliguid 2008).
12 May 2011	Members of the NPA disarmed security guards employed by St Augustine Gold and Copper Mining Limited who were accompanying geologists en route to the King King Copper Gold Project. One security guard resisted and was killed by the NPA (Lim 2011).

The opposition of local government units

A localized backlash against mining

One of the most significant forms of opposition to mining has come from the archipelago's local governments the province, the municipality and the *barangay* (a submunicipal level of local government akin to a village in a rural area or a neighborhood in an urban area).[2] There has been an emergence of a localized

---

2 In the Philippines, the regions are not a form of local government but are a territorial organization of the archipelago designed for the provision of national government services. There are also cities in the Philippines that are a form of local government which is, essentially, a variant of the municipality containing some of the taxation powers of a province. For the purposes of this book, "municipality" and "city" are treated as interchangeable terms.

backlash against mining where local governments have turned to local laws to ban mining from their territories (Hatcher 2010). Local governments "have become active in resisting mining entry and operations in their jurisdictions" (Lansang 2011, 154).

Local governments in the Philippines

During the Marcos presidency, political power was centralized to a much greater degree than it had ever been before (Holden and Jacobson 2006). One of President Corazon Aquino's first priorities after the "People Power" revolution of February 1986 was to institutionalize the empowerment of people at the local level and pave the way for local democratic development. To this end, the 1987 Constitution was specifically designed to amplify the powers of local governments in the Philippines, and Section 1 of Article X of the 1987 Constitution delineated the local governments as being the province, the municipality and the *barangay*. In 1991, Congress passed the Local Government Code – one of the most influential statutes ever passed in the islands – and it formalized these three types of local governments (Holden and Jacobson 2006).

Civil society access to local government units

One of the most profound aspects of the Local Government Code is the way it gives civil society organizations access to local governments (Holden and Jacobson 2006). There are provisions in the Local Government Code mandating NGO participation in local government[3] and authorizing the provision of financial assistance from local governments to NGOs.[4] A community affected by a mining project, such as Barangay Macambol, can approach an NGO and seek its assistance in opposing mining. The same community can then approach its local government and seek the local government's assistance in opposing mining as well. However, it is quite possible that by the time the community has approached the local government, the NGO it has also approached will be making use of the provisions in the Local Government Code – and will be working against mining through the local government's legislative process and may even be receiving funding from the local government to do this! Through the tripartite relationship between concerned community, NGO and local government, substantial and effective opposition to mining has been advanced (Holden and Jacobson 2006).

---

3 Sections 34 and 35.
4 Section 36.

The withholding of consent by local governments

The first way in which local governments have opposed mining is through the withholding of consent to mining operations (Holden and Jacobson 2006). There are provisions in the Local Government Code requiring national government agencies (such as the DENR in the case of mining projects) to consult with and obtain the consent of all local governments where a project is located prior to that project being undertaken.[5] If a mine is going to be located in a particular *barangay*, the DENR must obtain the consent of this *barangay*, as well as the consent of the municipality and province where it is located, before the mine can proceed (Holden and Jacobson 2006). Some local governments have passed resolutions opposing mining as a way of signaling that they have no intention of consenting to any mining projects within their jurisdiction. In 2008, the province of Palawan passed a resolution imposing a 25-year moratorium on small-scale mining activities[6] and declaring the province's resolve to oppose, firmly, any new applications for large-scale mining projects. In January 2011, the province of Negros Occidental passed a resolution[7] expressing its opposition to all existing and future mining applications in Negros Occidental, stating that for the next 25 years no favorable endorsements to any mining projects will be granted. In March of 2011, the province of Albay declared its strong opposition to any mining exploration and mining activity in the entire province of Albay.[8]

The withholding of consent by local governments is viewed as a serious impediment to the efforts of the national government to use mining as a vehicle for development. If local governments are refusing to consent to mining, it becomes substantially more difficult for the mining industry to conduct mine exploration and development in the Philippines. In 1999, the DENR attempted to minimize the ability of local governments to withhold consent by issuing DENR AO 1999–34 which in Section 8 stated that only two of the three local governments (*barangay*, municipality and province) need consent to a mining project (Holden and Jacobson 2006).[9] This would

---

5 Section 2 (c), Section 26 and Section 27.
6 Resolution No. 7728–2008. Many mining companies were obtaining small-scale mining permits, thus allowing them to circumvent the requirements of the EIA system and developing mines which were, for all intents and purposes, large-scale (Fonbuena 2008). To stop this practice on Palawan, the resolution was passed prohibiting small-scale mining.
7 Resolution No. 0055–2011.
8 Resolution No. 020–2011.
9 DENR AO 1999–34 was replaced by DENR Memorandum Order No. 2004–09 which did not require that only two of the three local governments need consent to a mining project, but instead required, in Section 5, that a majority of the concerned local governments need consent. Again, this is still a departure from the requirements of the Local Government Code that all affected local governments (*barangay*, municipality and province) need consent to a mining project.

mean that a mining project could proceed even if the *barangay* hosting the mine did not consent as long as the municipality and province provided their consents. This could allow a mine to proceed even if the local government in greatest proximity to that mine did not consent as long as two local governments, more remote from the mine, consented. This shows that those people most directly affected by the operation of the mine would not be required to provide their consent.

This administrative order is problematic for two reasons: first, the Local Government Code requires the consent of all affected local governments; second, this administrative order is a regulation issued by a government agency and thus cannot preempt the provisions of the Local Government Code, which is a statute passed by Congress (Holden and Jacobson 2006). To allow this would give the DENR legislative powers, and Section 1 of Article VI of the 1987 Constitution reserves legislative powers to the Congress.

Local government mining moratoriums

The second way in which local governments have opposed mining is by taking advantage of various provisions of the Local Government Code[10] and implementing moratoriums banning mining. Currently nine provincial governments have passed moratoriums banning large-scale mining within their jurisdictions (Figure 6.3).

In 1999 the province of Capiz declared a 15-year moratorium on all large-scale mining activities,[11] while its neighbor on the island of Panay (the province of Iloilo) banned large-scale mining for 15 years.[12] Three years later, the resolution in the province of Iloilo was amended extending the ban from 15 years to 50 years (Burgos 2011). In 2002, the province of Oriental Mindoro implemented a 25-year moratorium on all large-scale mining projects[13] and this resolution was confirmed again in 2008.[14] In 2003, the province of Samar imposed a 50-year large-scale mining moratorium[15] while its neighbor, the province of Eastern Samar, imposed an indefinite moratorium on the development of any new large-scale mines (subject to a

---

10 Sections 16 and 17 taken in conjunction with Sections 447(1) 447(5), 468(1), and 468(4).

11 Resolution No. 006–1999. The staff of the Environmental Legal Assistance Center drafted the Capiz mining moratorium thus demonstrating the close relationships between local governments and NGOs (Mayo-Anda 2005).

12 Resolution No. 145–1999.

13 Resolution No. 001–2002.

14 Resolution No. 313–2008.

15 Resolution No. 541–2003.

Figure 6.3. Provinces with mining moratoriums

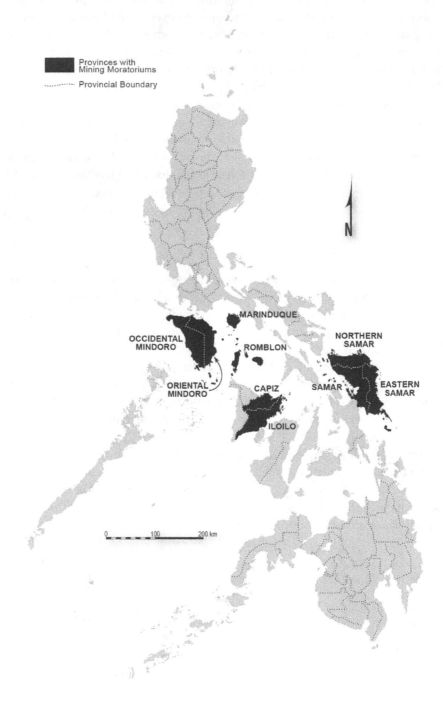

"grandfathering" provision allowing existing mining operations such as the Homonhon Chromite Project to continue operations).[16] That same year also saw Northern Samar pass a resolution opposing any mining operations or similar activities on Batag Island in the municipality of Laoang.[17] Four years later this was followed by a province wide mining moratorium banning all forms of large-scale mining from Northern Samar for 50 years (Labro 2010). In 2005 the province of Marinduque implemented a 50-year large-scale mining moratorium[18] and this was confirmed again in 2007.[19] In 2009, the province of Occidental Mindoro imposed a 25-year large-scale mining moratorium.[20] Lastly, in 2011, Romblon governor Eduardo Firmalo declared an indefinite moratorium on mining operations in the province of Romblon.[21]

While not a full scale moratorium precluding all large-scale mining, in June of 2010 the province of South Cotabato passed a resolution banning open-pit mining (Lansang 2011). Given the location of the Tampakan Project in South Cotabato, this vote is highly significant as it is clearly directed to stop this particular mine from being developed as the project proponent has made it clear that it intends to develop this mine using open-pit methods (Xstrata 2008).

According to Attorney Grizelda Mayo-Anda, the preponderance of mining moratoriums in the Visayan Islands is attributable to the fact that some of these provinces (namely Marinduque) have had highly unpleasant experiences with mining and due to the fact that they are islands – meaning they see mining's negative impacts much more quickly than they would if they were landlocked provinces in the interior of Luzon or Mindanao (Mayo-Anda 2009). As islands, they are highly dependent upon aquatic resources so they are vulnerable to mining-related water pollution; Capiz, for example is the "seafood capital of the Philippines" (Mayo-Anda 2009, interview). The moratoriums in the three provinces on the island of Samar are noteworthy in that they are examples of how the growing ecological consciousness of the Samarnons has moved them to oppose mining (Santos and Lagos 2004). Samar and Northern Samar are high typhoon risk locations while Eastern Samar is at medium risk of experiencing typhoons (Figure 4.1). All three provinces are at medium risk of experiencing earthquakes (Figure 4.2)

---

16 Resolution No. 008–2003.
17 Resolution No. 019–2003.
18 Resolution No. 379–2005.
19 Resolution No. 25–2007.
20 Resolution No. 140–2009.
21 Executive Order No. 001–2011.

and all three are at medium risk of experiencing *El Niño*–induced drought (Figure 4.11). The Samarnons are opposed to mining because of their vulnerability to typhoons and mining's lack of economic benefits for the poor (Mayo-Anda 2009).

The mining moratoriums, much like local government refusals of consent, are also controversial and are viewed in an unfavorable light by both the mining industry and the national government (Holden and Jacobson 2006). Michael Cabalda, chief science research specialist of the MGB, articulated a view that these moratoriums are "contrary to national policy" and that they represent a "lost opportunity" (Cabalda 2004, interview). The DENR has asked the Department of Justice (DOJ) and the Department of Interior and Local Government (DILG) for their respective opinions on the validity of local government mining moratoriums (Holden and Jacobson 2006). The DOJ has declined to provide an opinion stating that to do so would be to exercise a judicial function and to act in excess of its jurisdiction.[22] However, the DILG issued an opinion stating that local government mining moratoriums are invalid.[23] The Department of Interior and Local Government bases its opinion on a strict interpretation of portions of the Local Government Code[24] which state that local governments have the authority to pass laws pursuant to national policies and subject to supervision, control and review of the DENR pertaining to the enforcement of forestry laws limited to community-based forestry projects, pollution-control law, small-scale mining law and other laws on the protection of the environment. The DILG has placed an emphasis upon the requirement that such laws comply with national policy (the efforts of the Philippine government to encourage more mining) and upon what it sees as a confinement of the jurisdiction availed by this section to small-scale mining, in rendering its opinion that such moratoriums are invalid.

Attorney Mayo-Anda regards this as a moot debate. To Mayo-Anda, local governments acquire the jurisdiction to implement mining moratoriums from a broad reading of the Local Government Code as well as by using the provisions of the Wildlife Resources Conservation and Protection Act (Republic Act 9147) which empower local governments with the ability to declare areas wildlife habitat zones and thus exempt them from extractive activities such as mining (Mayo-Anda 2005). Given the provisions in the Local Government Code, encouraging close relations between civil society organizations and local governments and providing the ability to declare areas

---

22 DOJ Opinion No. 8, S. 2005.
23 DILG Opinion No. 152 S. 2003.
24 Section 17 (b) (3) (iii).

as wildlife habitat zones implements an ideal aperture for social movements to act against mining.

The Alternative Mining Bill

Progressive legislation in response to the risks

Perhaps the most tangible manifestation of how civil society groups in the Philippines reject the promises of mining as the new modernity is the Alternative Mining Bill. On 13 May 2009 a group of legislators in the House of Representatives of the Philippine Congress, which included two party-list legislators,[25] concerned with mining's environmental and social effects filed House Bill 6342, seeking to repeal the Mining Act of 1995 and to replace it with an Alternative Mining Act.[26] Work started on the Alternative Mining Bill almost immediately after the passage of the Alternative Mining Act in 1995 and this work began to accelerate in October of 2002 after the Dapitan Initiative when 24 leading civil society representatives attended a meeting in Dapitan, Zamboanga del Norte and signed an initiative calling for the repeal of the Mining Act, the cancellation of all mineral production agreements, and a national moratorium on the issuance of large-scale mining permits for 100 years (Lansang 2011). The Alternative Mining Bill was drafted by the Legal Rights and Natural Resources Center and was proposed in response to the rhetorical question frequently posed by mining's proponents to its opponents, "What is your alternative?" The Alternative Mining Bill was prepared to frame mining so as to more properly address development, and there are a number of provisions in this bill designed to ameliorate the environmental and social effects of mining; in particular, some of them are specifically designed to address the potential for mining-induced disasters caused by the intersection of mining with natural hazards (Obanil 2009).

---

25 Article VI of the 1987 Constitution and the Party-List System Act, Republic Act No. 7941, provided for the election of representatives from various traditionally marginalized sectors such as workers, peasants, the urban poor, indigenous peoples and the youth (Holden 2009a). Party-list representatives are entitled to 20 percent of all seats in the House of Representatives and the parties receive one seat for every two percent of the vote they receive, with additional seats being distributed in proportion to their total number of votes received, with each party to receive no more than three seats. Two of the sponsors of House Bill 6342 were Walden Bello and Risa Hontiveros-Basaquel, from the Akbayan Citizens' Action Party, a democratic socialist party.

26 House Bill 6342 lapsed with the end of the Fourteenth Congress in May of 2010 but when the Fifteenth Congress was elected, three new bills were filed which all contained provisions for an Alternative Mining Bill: House Bill 206, filed on 1 July 2010; House Bill 3763, filed on 1 December 2010; and House Bill 4315, filed on 2 March 2011.

A prohibition of mining in hazard-prone areas

Section 39 of the Alternative Mining Bill[27] prohibits mining in areas prone to climatic disasters (typhoons and El Niño–induced drought) and geological hazards (earthquakes, tsunamis and volcanoes). Rovik Obanil stated that Section 39 would prohibit mining in almost all of the Philippines since almost the entire archipelago is subject to some degree of risk from either climatic disasters or geological hazards (Obanil 2009). This has been done to implement safeguards protecting vulnerable communities (living in proximity to mines) from mining's adverse environmental effects. Jesus Garganera bluntly admitted that Section 39 would prohibit mining in most of the archipelago but quickly qualified this by stating that this is its intent (Garganera 2009). The objective of this section is to make sure that mining does not pose a threat to communities living where mines are located. To Engineer Virgilio Perdigon, Section 39 would prohibit mining in most of the Philippines; this has been done to undo the havoc inflicted on the Philippines by allowing mining amid the natural hazards prevalent there (Perdigon 2009).

A calamity protection fund

Section 108 of the Alternative Mining Bill[28] requires all mineral agreement holders to deposit PHP 5 million (approximately USD 115,000) a year into a fund maintained by the government to be used for redressing the effects of calamities relating to mining activities. According to Rovik Obanil, the requirement of a calamity protection fund in Section 108 is based on what should be best practices for mining (Obanil 2009). This has been designed to prevent unfunded environmental liabilities and to ensure that there will be a fund that can be used for rehabilitation. Obanil stated that an important influence on the Legal Rights and Natural Resources Center when it was drafting Section 108 were Miranda et al. (2005, 43) who wrote:

> Financial sureties are not generally required for catastrophic events such as earthquakes, floods, tailings dam failures, or the unanticipated onset of acid mine drainage after mine closure. Where such incidents have occurred the public has generally been responsible for a large part of the cleanup costs. A national fund or financial pool could be established to pay for catastrophic events.

---

27 These provisions were carried on into Section 40 of House Bill 206, Section 40 of House Bill 3763 and Section 37 of House Bill 4315.

28 These provisions were carried on into Section 110 of House Bill 206, Section 112 of House Bill 3763 and Section 108 of House Bill 4315. The financial costs of mine reclamation will be discussed in Chapter Seven.

The management of tailings dams

Section 114 of the Alternative Mining Bill[29] requires that tailings impoundments must not endanger critical watersheds or low-lying valleys. Much like Section 39, this will also prohibit mining in almost all of the Philippines but this is precisely the point behind the section as environmental protection should be thought about before mining; "environmental protection, human rights and food security should come before economic gain" (Obanil 2009, interview). Engineer Virgilio Perdigon also views this as something that would prohibit mining in large portions of the archipelago but there are many places where mining just should not occur (Perdigon 2009). To Ricardio Saturay, the requirement in Section 114 that tailings impoundments be built away from critical watersheds emanates from the fact that there have been several instances where tailings dams have failed in such areas in the Philippines (Saturay 2009).

Section 114 of the Alternative Mining Bill also requires that tailings dams comply with the international standards for large dams. "The Philippines," wrote Carreon (2009, 107), "is among the worst countries in the world [for] tailings dam failures." In Rovik Obanil's view, what happened at the Marcopper Mine on the island of Marinduque in 1996 is an example of how tailings dams in the Philippines do not comply with the international standards for large dams (Obanil 2009). Engineer Virgilio Perdigon articulated a view that, currently, tailings dams in the Philippines do not meet the international standards for large dams (Perdigon 2009). Jesus Garganera agreed with both Obanil and Perdigon stating that there is substantial anecdotal evidence that tailings dams in the archipelago do not comply with the international standards for large dams – and in any event, the DENR lacks the necessary resources to properly monitor tailings dams (Garganera 2009). Dr Emelina Regis does not regard tailings dams in the Philippines as currently complying with the international standards for large dams and wonders whether even these would be adequate given the unique set of risks faced in the archipelago (Regis 2009).

From Risk Society to the Viability of Mining

Chapter Five made it clear that those opposed to mining in the Philippines do not view technology as a solution to the risks created by locating large-scale mining projects among the natural hazards present in the archipelago. The view that disasters are an aberration or departure from normality

---

29 These provisions were carried on into Section 116 of House Bill 206, Section 118 of House Bill 3763 and Section 114 of House Bill 4315.

capable of being overcome by technocratic solutions is emblematic of faith in expertly engineered progress – a hallmark of modernity. Modernity is, however, a project that has not achieved full acceptance in the Philippines. Locating a high risk and high consequence activity (such as hardrock mining) among vulnerable populations who are vitally dependent upon access to natural resources for subsistence constitutes an example of risk society. Clearly it is a situation where the biggest threats to people come, not from the forces of nature, but from technology. It is in response to the risks arising from mining amid hazards that substantial opposition to mining has emerged. This opposition has involved protests (Figure 6.2), litigation, administrative proceedings and actions undertaken by the powerful Roman Catholic Church – an institution with a firm commitment to the poor and marginalized. On some occasions this opposition has even included armed attacks on mining projects being carried out by the NPA (Table 6.1). Two noteworthy aspects of this activism against mining have been the passage of moratoriums banning mining by nine provincial governments (Figure 6.3) and the filing of a bill in the House of Representatives that would prohibit mining in areas subject to climatic and geological hazards.

Attention now turns to a discussion of whether mining may not be an appropriate prescription for achieving economic growth in the Philippines. Will mining generate a tremendous flood of prosperity that lifts all boats and more than compensates for its environmental harm? Or, alternatively, will mining generate lasting environmental harm and produce benefits which are so inequitably shared among the members of society that little, if any, benefit accrues to the poor and marginalized?

Chapter Seven

MINING AS A FLAWED DEVELOPMENT PARADIGM

Mining: A Questionable Development Strategy

Given the risks associated with locating large-scale mining projects amid the natural hazards present in the Philippines, and the reluctance of many members of that archipelago's civil society to accept technological solutions to these risks, one may wonder whether a mining-based development paradigm is an appropriate approach to be followed. Will mining-related environmental disruptions brought on by the interactions of mining's environmental effects and the natural hazards present in the Philippines only serve to disrupt the ecology of the poor and end up impoverishing vulnerable communities adjacent to mining operations? Alternatively, will mining act as an engine of economic growth and generate so much prosperity that whatever instances of environmental disruption may occur can easily be compensated for by the subsequent rising tide of prosperity that "lifts all boats?"

The Twin Pillars of Sustainable Development

In addressing the efficacy of any development strategy, the concept of sustainable development is a useful metric. In 1987, the World Commission on Environment and Development defined the now ubiquitous term "sustainable development" as being "development that meets the needs of the present without compromising the ability of future generations to meet their own needs" (World Commission on Environment and Development 1987, 43). However, this definition may only be viewed as a starting point in discussions of sustainable development because much of this discussion directs its attention not on the negative consequences of economic growth upon the environment, but on the negative consequences of environmental degradation upon economic growth (Holden 2009b). Many of those calling for sustainable development place their priority upon sustaining growth instead of placing their priority upon sustaining the environment (Escobar 1996). Indeed, there are some who adopt a view that poverty itself is the cause of environmental degradation and that poor people are those who are despoiling the environment. This controversial view holds that growth is needed to eliminate poverty and only if poverty can be eliminated can protection of the environment even begin to be discussed. The World Commission on Environment and Development (1987, 28) was guilty of this when it stated: "Those who are poor and hungry will often destroy their immediate environment in order to survive. They will cut down forests; their livestock will overgraze grasslands; they will overuse marginal land; and in growing numbers they will crowd into congested cities." A more robust discussion of sustainable development requires an assessment of two important concepts: intergenerational equity and intragenerational equity (Table 7.1).

For something to be sustainable, it must comply with the notion of intergenerational equity and cannot benefit present generations at the expense of future generations (George 1999; Holden 2009b; Martin 2003). This theme of intergenerational equity is what is most apparent from a reading of the World Commission on Environment and Development and it may be thought of as a necessary condition for sustainability. For something to be development, however, it must comply with the notion of intragenerational equity; it must have benefits that are equitably shared among members of the current generation (George 1999; Holden 2009b; Martin 2003). Every development improves someone's quality of life, even if it is only the developer's. For something truly to bring development, however, there must be benefits that are widely spread among members of the current generation. Therefore, this theme of intragenerational equity may be thought of as a necessary condition for development. With these two necessary conditions in mind, an argument may be made that a development paradigm based upon the promotion of large-scale mining does not generate sustainable development for three reasons: first, mining does not benefit current generations; second, mining imposes costs on future generations; third, mining does not generate widespread benefits.

Table 7.1. The twin pillars of sustainable development

Concept	What this entails	Necessary condition for
Intergenerational equity	Development must not benefit present generations at the expense of future generations	Sustainability
Intragenerational equity	Development must have benefits which are equitably shared among members of the current generation	Development

Source: George (1999), Holden (2009b) and Martin (2003).

Mining: A Lack of Benefits to Current Generations

Mining: A poor source of employment creation

The first problem with relying upon mining as an agent of economic development is that modern hardrock mining is a capital-intensive industry, not a labor-intensive industry; accordingly, it does not create a large number of jobs (Power 1996, 2002, 2005, 2008). In 2001 mining was responsible for only 0.3 percent of all employment in the Philippines (Carreon 2009). By 2009, notwithstanding the aggressive efforts of the government to promote mining over the intervening eight years, this number grew to only 0.5 percent of all employment in the archipelago (Joint Foreign Chambers of the Philippines 2010). When one compares this to agriculture, the source of two out of every three jobs in the Philippines, perspective is gained on the paltry prospects for employment creation availed by mining. As Hatcher (2010, 20) wrote about mining in the islands, "A number of studies on employment rates and the mining sector suggest that the industry's potential for job creation has repeatedly been overstated."

This lack of employment opportunities becomes even more of an issue when one considers the limited scope for employment creation at the scale of the communities directly affected by mining projects. At the Rapu-Rapu Polymetallic Project in 2006, there were 286 people employed directly at the mine; out of these 286 people, 93 people were people who lived on Rapu-Rapu Island while the remaining 193 employees were people who had come to work at the mine from elsewhere in the Philippines and from abroad (Oxfam Australia 2008). This means that out of the 9,557 people living on Rapu-Rapu Island, the mine employed less than one percent of all islanders. The communities in immediate spatial proximity to mining projects are the people most vulnerable to any environmental disruption caused by the mine. Quite often though, as demonstrated by the Rapu-Rapu Polymetallic Project, these people receive little direct benefit from the mining projects in terms of employment opportunities. This puts these people in the unenviable position of being in a space of vulnerability but not a space of benefit.

Often, as discussed in Chapter Four, mining projects are located on lands inhabited by indigenous peoples – the most marginalized members of the archipelago's society. The government maintains that "indigenous peoples can be the main beneficiaries of large mining operations" (Neri 2005, 25). This is a problematic assertion to make as it assumes that some of the poorest and least educated members of society will become employed in a capital-intensive industry that uses highly sophisticated technology. Consider the Mount Apo geothermal project on the island of Mindanao (a capital-intensive activity much like a modern hardrock mine). When the geothermal project was designed, the Philippine National Oil Company promised the indigenous Manobo inhabitants residing in the vicinity of Mount Apo that they would be the first ones to be hired; however, when the project became operational, they turned out to be the first ones to be fired because they lacked the necessary technical skills (Alejo 2000).

Mining: A low scope for tax revenues

The second problem with relying on mining to act as a vehicle for economic development is the low scope for tax revenue under the Mining Act of 1995. As indicated in Chapter Two, the Mining Act contains a number of generous incentives to encourage mining, such as a four-year income tax holiday, tax and duty-free capital equipment imports, value added tax exemptions, income tax deductions where operations are posting losses and accelerated depreciation. While these generous incentives encourage the entry of mining companies into the Philippines by assuring that they will have to pay little or no taxes, these incentives serve to have the effect of dramatically reducing the scope this industry provides to act as a source of revenue for the state. In 2008, for example, the mining industry was responsible for less than one percent of all government revenue and many feel that the fiscal regime under the Mining Act is "heavily stacked against the state and in favor of the mining companies" (Landingin 2008).

A lack of linkages to other industries

The third problem with relying on mining as an agent of economic development is that mining has few linkages to other industries. If mining does not create a lot of jobs on its own it could still act as a propulsive force driving the economy if it has substantial linkages to other industries within the country. If an industry has substantial backward linkages to other industries within a nation, it will be purchasing large amounts of its inputs from those other industries and will have a high potential to act as an engine for economic development. If an industry has substantial forward linkages to other industries within a nation, it will be selling large amounts of its outputs to other industries and will also have a high potential to act as a force stimulating economic development.

In the Philippines there are few – if any – backward linkages from the mining industry to other industries within the Philippines. In Table 7.2, some examples of mining equipment and mining equipment suppliers are displayed, and all of this equipment comes from corporations in other countries. This severely limits the backward linkages from the mining industry to other industries within the archipelago and this will weaken the ability of the mining industry to stimulate the economy. "The mining industry imports almost all

Table 7.2. Mining equipment and its suppliers

Mining equipment	Major supplier
Motor controls	Siemens (Germany)
Tools	Kumoto (Japan)
Chemicals	Philips (Holland)
Loaders and earthmovers	John Deere (USA); Caterpillar (USA, Canada); Terex (Germany)
Flotation equipment	Sandvik and Association; Swedish Steels (Sweden)
Steel casting and alloys	ESSO (USA)

Source: Tujan and Guzman (2002).

of its capital equipment and intermediate inputs such as fuel, transportation vehicles, generators, drills, bulldozers, explosives and chemicals used for initial processing" (Tujan and Guzman 2002, 61).

Likewise, the integration of the mining industry into other industries within the Philippines is very poor. The Philippine mining industry operates as a predominantly isolated entity (Tujan and Guzman 2002). This exemplifies how mining complexes often take the form of enclave economies, developing relatively few links to local suppliers or customers (Bebbington et al. 2008). The government has a "weak resolve" to "promote metal-based value added industrialization" (Ofreneo 2009, 203). This lack of resolve on the behalf of the government to promote a vertically integrated metals industry can be attributed to the domination of Philippine society by an oligarchy that has historically derived its powers from the production and export of unprocessed primary sector products (Hawes 1987). These people saw little, if any, need for industrialization based on their products and consequently, they developed a disdain for industrialization in general. Since this elite was also dominant politically, this generated little domestic incentive or interest in the industrialization of the country by the state.

The lack of forward linkages from gold mining is a substantial source of controversy seeing as between 80 to 85 percent of all gold mined in the world is used to make jewelry with much of this being used as wedding dowries in India (Sampat 2003). Only around 12 percent of all gold mined is used for industrial purposes, such as electronics (Young 2000). With gold being produced at 61 percent of the mines outlined in Table 2.1, (and with gold being the only mineral produced at 44 percent of these mines) it is difficult to envisage an extensive amount of forward linkages flowing from mining to other industries located within the Philippines. There are electronics assembly plants in the archipelago's export processing zones, but these overwhelmingly perform labor-intensive assembly functions involving minimal amounts of manufacturing whereby raw materials are fabricated into unassembled parts (Tujan 2007).

One thing that has greatly changed the spatial distribution of forward linkages from mining to other industries has been the dramatic reduction of transportation costs (Pegg 2006; Power 2002). In the late nineteenth and early twentieth centuries, high transportation costs necessitated the processing of raw materials in close spatial proximity to their sources; this caused the smelting of minerals to take place near mines. With the smelting of many minerals taking place near mines, mining would generate forward linkages to other industries in the immediate vicinity. Today, however, low transportation costs no longer necessitate that raw materials be processed in close spatial proximity to their sources. "Copper mined in Chile gets smelted in Europe and may end up in radiators of cars made in Japan and driven in California" (Sampat 2003, 112). Since low transportation costs have eliminated the spatial association between mining and smelting, mining now generates substantially fewer forward linkages to other industries in the immediate vicinity of mines and this has further severed the potential for mining to act as a growth pole stimulating economic development.

The volatility of mineral prices

The fourth problem with relying on mining as an agent of economic development is that mineral prices are notoriously volatile (Bebbington et al. 2008; Bridge 2004; Clements 1996; Lima and Suslick 2006; Power 2002, 2005, 2008). Mineral prices display substantial instability over time, as seen in Figures 7.1, 7.2 and 7.3 which display gold, copper and nickel prices in the United States (in 1998 constant dollars) from 1900 to 2009.

If a mine is planned and mineral prices suddenly rise, this will provide a windfall dividend for the project proponent as its estimates of profitability will be underestimated; however, if a mine is planned and mineral prices suddenly fall, this could prove catastrophic for the proponent as the declining prices could destroy the profitability of the investment before the first ton of ore has been mined. What causes the volatility of mineral prices is the fact that the major consumer of minerals is the manufacturing sector and it is heavily influenced by changes in macroeconomic conditions (Peck et al. 1992). In 1970, for example, there was a 0.3 percent drop in the gross national product of the United States; this led to a 3.3 percent decline in the sales of consumer durables (such as refrigerators); this, in turn, led to a 12 percent decline in demand for copper, which ultimately led to a 34 percent decline in the copper industry's overall profitability (Soussan 1988). Minute changes

Figure 7.1. Real gold prices in the United States, 1900–2009

Source: United States Geological Survey (2011).

Figure 7.2. Real copper prices in the United States, 1900–2009

Source: United States Geological Survey (2011).

in the demand for manufactured products (such as automobiles, toasters, refrigerators, washing machines and dryers) can generate changes in the quantity demanded of minerals that are amplified by orders of magnitude. Should a mine come into production during a period of protracted recession (such as that occurring in the early 1980s), its viability will be threatened by the level of demand for the ore it produces. This volatility of mineral prices is one of the main reasons why economists have long identified mining as

Figure 7.3. Real nickel prices in the United States, 1900–2009

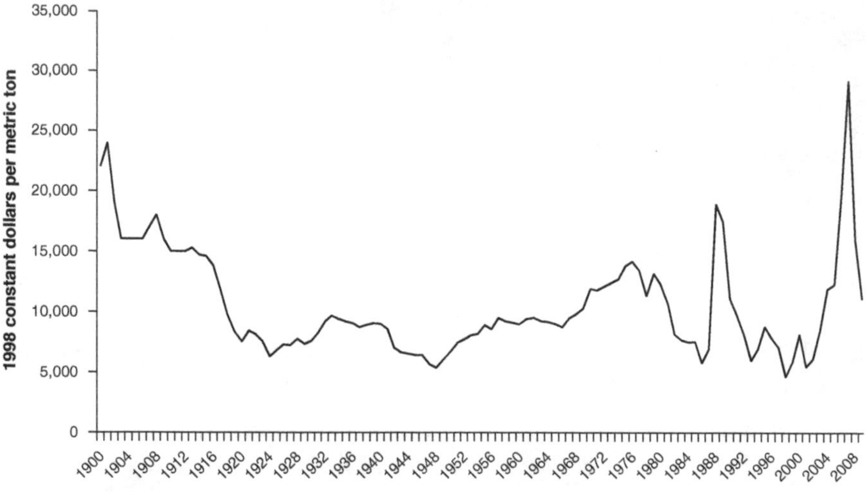

Source: United States Geological Survey (2011).

an industry with a high bankruptcy rate. As long ago as 1776, Adam Smith wrote the following in the seminal book *The Wealth of Nations*: "Of all those expensive and uncertain projects, however, which bring bankruptcy upon the greater part of the people who engage in them, there is none perhaps more perfectly ruinous than the search after new silver and gold mines" ([1776] 1991, 500). By aggressively advocating mining as a vehicle for development, the government is making the economy of the archipelago vulnerable to crises occurring in international markets. Windfall profits will be earned by mining companies during periods of high prices but conversely, the economy will be thrown into disarray when prices suddenly fall. Such a development strategy is not sustainable; as the World Commission on Environment and Development (1987, 53) stated, "Economic development is unsustainable if it increases vulnerability to crises."

The long-term downward trend of real mineral prices

The falling real prices of minerals over time

In addition to being volatile, many mineral prices have also experienced a long-term downward trend (Crowson 2007; Radetzki et al. 2008). In Figures 7.4 and 7.5, time-trends are inserted into the graphs of copper prices and nickel prices in the United States (in 1998 constant dollars) from 1900 to 2009. From 1900 to 2009, real copper prices fell by 0.5 percent a year on average, and real nickel prices fell by 0.63 percent a year on average (United States Geological Survey 2011).

Figure 7.4. Time trend of real copper prices in the United States, 1900–2009

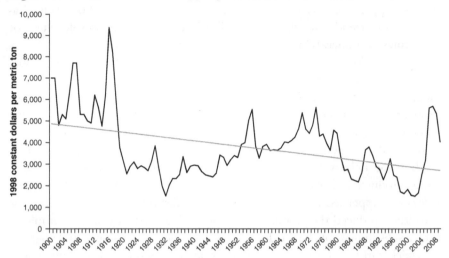

Source: United States Geological Survey (2011).

Figure 7.5. Time trend of real nickel prices in the United States, 1900–2009

Source: United States Geological Survey (2011).

The phenomenon of dematerialization

Over time there has been a gradual trend in manufacturing to consistently use less raw materials per unit of output and by the year 2000 the amount of raw materials per unit of industrial output was only 40 percent of what it was in 1930 (De Rivero 2001). In the twenty-first century, this dematerializing trend in metals will have an increasing impact upon countries extracting minerals such as copper, iron ore, lead, nickel, tin and zinc (De Rivero 2001). What has initiated this trend towards dematerialization has been the development of substitutes for minerals brought about by periods of high prices.

Consider, for example, the price of copper. From 1932 to 1956 copper prices rose due to the recovery of the global economy from the Great Depression, the production of armaments for the Second World War, the postwar reconstruction of Japan and Western Europe and the Korean War (Lopez 1992). Then from 1958 to 1970 copper rose again due to the Vietnam War and due to political disturbances in major copper producing countries (such as the Belgian Congo, Chile and Zambia), which created shortages in world production (Lopez 1992). During both of these time periods, the high prices of copper led to the development of substitutes for copper and these substitutes reduced the amount of copper being used. A classic example of such a substitute would be the development of fiber-optic cables – 40 kilograms of which can transport as many telephone calls as one ton of copper wire (De Rivero 2001). At the Los Alamos National Laboratory in New Mexico, scientists have prepared a new superconducting tape that can carry 1,200 times the amount of electricity as that carried by a copper cable (De Rivero 2001). In short, "the demand for raw materials is declining over the long run" (De Rivero 2001, 101). For the government of the Philippines to place a heavy reliance upon mining as an economic development paradigm in an age of dematerialization is to court disaster. Reliance is placed upon revenue obtained from extracting minerals but over time, the real prices of these minerals are progressively declining so more and more of them must be extracted for them to provide a given amount of revenue.

The recent rise in commodity prices

One thing which may generate optimism for those who advocate mining as a vehicle for development is the upward trend in many commodity prices that has prevailed since 2002 (Crowson 2007; Rosenau-Tornow et al. 2009). In Figures 7.2 and 7.3 for example, copper prices rose steadily from 2002 until 2007 and nickel prices rose steadily from 2001 until 2007 (then both commodities tailed off somewhat with the global recession of 2008). This sudden increase in commodity prices has been caused by the insatiable demand for minerals in China and the lag in the mining industry's response to this demand (Tilton and Lagos 2007). The natural resource economics literature is clear: this is a short-term phenomenon and the world is not entering a period of high long-term commodity prices (Crowson 2007; Radetzki et al. 2008; Rosenau-Tornow et al. 2009; Tilton and Lagos 2007). Given enough

Figure 7.6. Scrap metal collectors, Davao City, January 2006

time, the mining industry will locate and develop more mines and increased commodity production will reverse the recent rise in commodity prices and the long-term downward trend will reassert itself. As Rosenau-Tornow et al. (2009, 161) wrote, "Our planet is large enough and only to a minimum explored, it still bears many hidden mineral deposits."

Another dynamic that may also off-set the recent upward trend in mineral prices is recycling; should prices continue to rise, more recycled metal products will be collected and this will provide an alternative source of minerals that does not have to come from the mining of virgin ore deposits (Figure 7.6). In the United States, for example, it is estimated that there are 40 million tons of copper sitting in landfills (Sampat 2003). With a mere 13 percent of all worldwide copper consumption coming from recycled sources, there is a substantial potential for recycled copper to ameliorate the recent increase in demand (Sampat 2003).

The recent rise in gold prices

The price of gold is different from the prices of base metals such as copper and nickel, as the price of gold is largely determined by its role as a store of value. "Gold is an industry that booms when times are bad elsewhere, and droops when other industries flourish" (Lewis 2010, 14). Unlike base metals, "gold does not need a lively manufacturing sector to be useful"; it only needs "untrustworthy currency and a mercurial stock market" (Lewis 2010, 14). For much of the twentieth century, the price of gold was fixed at a price such that central banks could approach the United States Federal Reserve and receive one ounce of gold for a set amount of dollars. This gold standard was suspended by President Franklin D. Roosevelt on 19 April 1933 due to the turmoil of the Great Depression, but was restored in 1944 and until its final abolition by President Richard M. Nixon on 15 August 1973, one ounce of gold was fixed at USD 35 (Hammes and Wills 2005). After the abolition of the gold standard, gold became a commodity subject to price fluctuations based upon changes in its supply and demand. From 1970 until 1980, the price of gold appreciated substantially (Figure 7.1) as it was perceived as a hedge against the high inflation rates prevailing during that decade. Then, from 1980 until 2001, the price of gold experienced a steady decrease as inflation was brought under control (Bordo et al. 2007). Since 2001, however, uncertainties regarding expectations of future inflation brought about by high levels of sovereign debt have caused an upsurge in gold prices from under USD 300 per ounce in 2001 up to USD 1,110 per ounce in 2009 (Lewis 2010). Should these concerns prove to be unfounded, gold prices could well tumble as they did from 1980 to 2002. Alternatively, if inflationary expectations are not addressed but indexed financial instruments are made widely available, gold could find itself confronted with a substitute in the market for hedges against inflation and this could lead to a fall in its price (Young 2000). Also, should there be a sell-off of central bank gold reserves this could generate a sudden decrease in the price of gold. Governments in the industrialized world hold substantial reserves of gold; by holding these reserves and keeping what amounts to a vast supply of gold off the market, prices are kept artificially high (Young 2000). Even if only a portion of this gold were to be put on the market, there could be a substantial decrease in the commodity's price.

The resource curse thesis

The crowding out of other economic sectors

One of the most problematic aspects of relying on mining as an agent of economic development is the resource curse thesis, where counter intuitively, mineral rich countries actually have poorer economic performance than those countries with less minerals (Auty 1994; Bebbington et al. 2008; Guenther

2008; Pegg 2006; Ross 2001; Sachs and Warner 1999, 2001; Stevens 2003; Stinjns 2005; Weber-Fahr 2002). In summarizing the resource curse thesis, Sachs and Warner (2001, 828) pose the question, "If natural resources really do help development, why do we [see no] positive correlation today between natural wealth and other kinds of economic wealth?" While complex, most current explanations of resource curse thesis hinge upon a crowding-out logic (Sachs and Warner 2001). In these explanations, the growth of a mining sector will crowd out other types of economic activity and the economy will end up becoming poorer, not richer.

The phenomenon known as the Dutch disease

One example of the most powerful types of crowding out that can occur is the phenomenon known as the Dutch disease (Auty 1994; Bebbington et al. 2008; Ross 2001). The term "Dutch disease" was coined after the appreciation of the Dutch currency – the gilder – during the development of North Sea oil in Holland during the 1970s. When North Sea oil revenues began flowing into the Netherlands, the gilder appreciated in value and this made it more difficult for the Dutch to sell their traditional exports in foreign markets. An extensive development of mineral resources during a period of high mineral prices can cause an extensive appreciation of a country's currency and this can crowd out all other types of economic activity by making other exports prohibitively expensive in foreign markets. Then when mineral prices fall, the country will find itself locked into the extraction of minerals with no alternative sources of economic activity.

Dutch disease and overseas Filipino workers

Dutch disease can pose a serious challenge to the Philippines because of its heavy reliance upon overseas Filipino workers (OFWs) who remit funds home to their families. An important component of overseas employment for many OFWs is the fact that employment abroad allows Filipinos to earn incomes far greater than would be possible at home, in large part because the peso is so devalued (Rodriguez 2010). As Rodriguez (2010, 32) wrote:

> The promise of earnings in foreign currencies stronger than the Philippine peso, however temporarily, becomes a draw for prospective migrants even if it slots them into lower paying jobs relative to native workers. Meanwhile, when workers' wages make their way back to the Philippines, the state is able to strengthen its foreign currency reserves and thereby pay its numerous debts.

There are many mines in the Philippines such as the Tampakan Project (one of the largest undeveloped copper and gold deposits in Southeast Asia), which are only in the developmental stage. Once these mines become fully developed and produce hundreds of thousands of ounces of gold over several decades, there will be upward pressure on the peso – particularly if gold prices remain high. When mining leads to an appreciation of the peso, this will mean that the remittances of OFWs (once converted back into pesos by their family members) will be reduced. These people will then be forced to work even longer hours and spend even and more time away from their families. This shows how one of the most important neoliberal polices of the government – the promotion of mining – operates orthogonally with another – the export of labor.

Mining and the Moro Islamic Liberation Front

A mining-based development paradigm could pose serious problems for the peace process between the Moro Islamic Liberation Front (MILF) and the government of the Philippines (Holden and Jacobson 2007b). The MILF are engaged in negotiations with the government to resolve the conflict that both sides have been engaged in since the early 1980s and it is desirous of achieving an independent Islamic state in the four Muslim dominated provinces of the Autonomous Region of Muslim Mindanao (ARMM). As Von Al Haq, the chair of the MILF Coordinating Committee on the Cessation of Hostilities, asked, "If the government of Indonesia, a Muslim country, can grant independence to East Timor, an area of its territory inhabited by Christians, then why cannot the government of the Philippines, a Christian country, grant independence to an area of its territory inhabited by inhabited by Muslims?" (Al Haq 2005, interview).

There is a view in the resources and conflict literature that resource abundance can lengthen a separatist conflict, such as that between the MILF and the Philippine government (Le Billon 2004, 2005). The lengthening may occur in either, or both, of two ways: first, the government may negotiate a peace accord with separatist rebels, and then renege upon it in order to gain access to resources; second, the separatists, in advance, may expect the government to renege upon the agreement and become hesitant to sign a peace accord. There are some in the Philippines who suggest that the Philippine government is insincere in its dealings with the MILF because it desirous of having access to resources on lands occupied by them. "One prevailing perception now is that the military operations against the MILF are part of the government plan to drive away the Muslims from rich areas that the government will give to investors – Filipino and foreign – for development" (Diaz 2003, 165). In the specific case of hardrock mining, it does not appear that there is a high degree of mineralization in the portion of Mindanao where the MILF operate (Holden and Jacobson 2007b). However, one must temper this by taking into account the fact that much of Mindanao's geology is relatively unexplored. It could be that the Philippine government is reluctant to negotiate a peace accord with the MILF, which would cede territory to an independent Muslim state because there is uncertainty over the extent of the resources that would be surrendered by such an accord. This uncertainty is certainly inimical to a resolution of the conflict between the parties.

In any event, regardless of the implications mining may pose for the peace process, it should be made clear that the MILF has gone on record declaring its opposition to mining by foreign corporations (Fernandez 2006). Jun Mantawil, chair of the MILF peace panel secretariat, declared that "The mining industry rapes, extracts, denudes, divests, drains and brings forth death and destruction to our environment and the marginalized population" (Fernandez 2006). During February 2008, the MILF asked the government to halt all mining applications within the proposed Bangsamoro territory, mostly within the ARMM (Landingin 2008). According to Al Haq, "the MILF believes that mining managed by multinational corporations is not good, multinational corporations can help the Bangsamoro people by not operating in the area. The Bangsamoro want to shape their destiny and they do not want foreign corporations to do it for them" (Al Haq 2005, interview). Al Haq also added a stark warning of the consequences befalling an attempt to pursue mining in territory claimed by the MILF without its consent by stating, "we will be able to manage the situation" (Al Haq 2005, interview).

Mining and the New People's Army

The extent of the overlap

The most widespread and serious threat to the security of the Philippine state is the New People's Army (NPA), the armed wing of the Communist Party of the Philippines (CPP). The spatial distribution of encounters between the AFP and the NPA from 2000 to 2004 was depicted in Figure 1.6 and this demonstrates the widespread presence of NPA activity across the archipelago. The locations of the major operating and proposed metallic mines from Figure 2.2 are overlaid upon Figure 1.6 and displayed in Figure 7.7. This shows that there is a substantial overlap between mining and NPA activity, particularly in northeastern and southeastern Mindanao. Consider, for

Figure 7.7. Confrontations between the AFP and the NPA and the location of mining projects

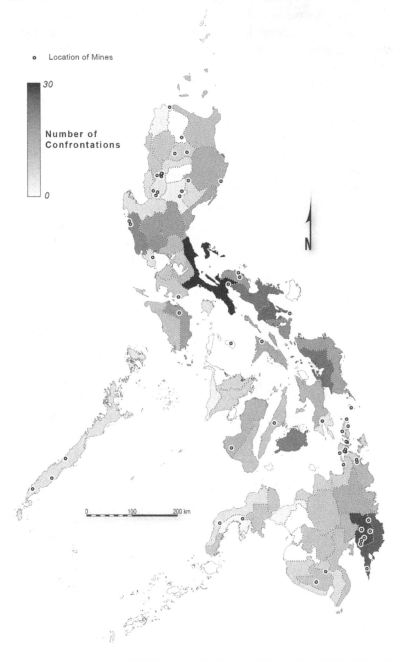

Source: Based on data from IBON (2001a, 2001b, 2002a, 2002b, 2003a, 2003b, 2004a, 2004b, 2005) and the Mines and Geosciences Bureau (2006, 2007a, 2007b, 2007c).

example, the province of Compostela Valley on the island of Mindanao. With less than one percent of the archipelago's land area, this province hosts 11 percent of all mining projects and was responsible for 5 percent of all confrontations between the AFP and the NPA from 2000 to 2004.[1]

There is a substantial overlap between mining and NPA activity because both are found in mountainous areas. Mines are located in mountainous areas (Figure 4.13) due to the orogenic process leading to mineralization. Mountainous areas also make ideal areas of operation for insurgent forces. "The role of geography," wrote Galula (1964, 23), "may be overriding in a revolutionary war." Rugged terrain helps insurgents because it is harder for counterinsurgents to control mountainous areas. According to the United States Army and United States Marine Corps (2007, 306), "Mountains, caves jungles, forests, swamps and other complex terrain are potential bases of operation for insurgents."

It is well-established Maoist doctrine to operate in mountainous areas. Maoists adhere to a doctrine known as the protracted people's war from the countryside that moves through three phases (Table 7.3) whereby the guerrillas gradually take control of the countryside and then surround and eventually take over the cities and, ultimately, destroy the state.

Table 7.3. The three phases of the protracted people's war from the countryside

Phase	What this entails
Strategic defensive	The government has a stronger correlation of forces and insurgents must concentrate on survival and building support
Strategic stalemate	Government and insurgent forces approach equilibrium and guerrilla warfare becomes the most important activity
Strategic counteroffensive	Insurgents have superior strength and military force and are able to engage in conventional operations to destroy the government's military capability

Source: United States Army and United States Marine Corps (2007).

---

1 There would be an even greater overlap between mining and NPA activity were it not for the mining moratoriums in all three of the provinces on the island of Samar, which by prohibiting large-scale mining, would reduce the extent of overlap between mining and NPA activity. This island has always had a substantial amount of NPA activity. From 2000–4, 8.39 percent of all confrontations between the AFP and the NPA in the Philippines occurred on this island even though it constitutes only 4.48 percent of the archipelago's land area.

The crucial component in the protracted people's war from the countryside – particularly in the initial strategic defensive phase – is what Mao Zedong called "base areas." According to Mao ([1937] 2005, 107):

> A guerrilla base may be defined as an area, strategically located, in which the guerrillas can carry out their duties of training, self-preservation and development. Ability to fight a war without a rear area is a fundamental characteristic of guerrilla action, but this does not mean that guerrillas can exist and function over a long period of time without the development of base areas.

To Mao Zedong, the impediments to movement imposed by rough terrain upon a superior state force that is attempting to suppress a guerrilla movement make the acquisition of guerrilla base areas in mountainous terrain essential. As it is well-established Maoist doctrine to operate in mountainous areas, the NPA (adherents to the protracted people's war from the countryside) are firmly committed to operations in mountainous terrain. In 2006 an article in *Ang Bayan*, the newsletter of the CPP, described the advantages of mountains stating, "Mountains provide effective concealment for snipers. A team or groups of NPA snipers can effectively pin down a company of enemy troops" (*Ang Bayan* 2006, 5). In mountainous areas, "the harshness and breadth of the terrain make it difficult for the enemy to trap NPA units that time and again concentrate and disperse" (*Ang Bayan* 2006, 6). This use of mountainous terrain by the NPA follows a long tradition in Philippine history; "One should also note that 'taking to the hills' had become part of the mass tradition of the oppressed dating to the days of the *remontados* of Spanish times" (Constantino 1975, 257).

The NPA also prefers to operate in mountainous areas because mountains usually constitute the boundaries of provinces in the Philippines (Caouette 2004). "Provincial border regions in the country are usually characterized by mountainous and forested terrain, with isolated *barangays* serving as ideal grounds for conducting guerrilla warfare" (Corpus 1989, 34). This is significant because border areas are where the state exhibits "bureaucratic ineptitude" (Kalyvas 2006, 133) or what may be called "areas of weak or confused political authority" (McColl 1969, 617). The Philippine National Police (PNP) is organized into separate provincial commands and when pursued by the PNP in one province, NPA members can cross into a different province and leave that PNP command's area of responsibility. Operating in border areas also allows a single guerrilla unit to exert influence on two or more provinces at once (Corpus 1989).

Revolutionary taxation of mining companies

One of the most controversial and problematic aspects of the interface of large-scale corporate mining with the NPA is the allegation that the NPA is engaging in the process of revolutionary taxation and is extorting money from mining companies. "Money," wrote Galula (1964, 40), "is the sinew of war" and the AFP has alleged that the NPA supports itself by collecting approximately USD 1.6 million a month in revolutionary taxes from logging firms, mining companies and plantations across the archipelago (Hastings and Mortela 2008). An unnamed foreign mining company executive told Landingin, "The NPA has engineers who look at your operations. They go to [the MGB website] and get the mineral deposit of the area and, from there, they calculate. They also go to the company website. Then, they compute." (2008, 8). "A mining company that fails to pay revolutionary taxes may find its facilities attacked, equipment destroyed and personnel killed" (Vivoda 2008, 134). The issue of whether revolutionary taxes will be imposed on large-scale mining projects is controversial because the NPA has traditionally denied levying revolutionary taxes on mining companies. In a January 2003 issue of *Ang Bayan*, the CPP discussed revolutionary taxation describing it as "a legitimate act and a right of the state of the people's democratic government" (*Ang Bayan* 2003, 3). "The money and resources collected from revolutionary taxation are used to defray the operations of the Party, the NPA, the people's democratic government and the revolutionary movement as well as projects for the general welfare of the people" (*Ang Bayan* 2003, 4). Nevertheless, the same article stated that revolutionary taxes would not be levied upon activities that generate environmental harm such as mining companies. As the article, (*Ang Bayan* 2003, 5), stated: "In implementing revolutionary taxation, the interest and welfare of the people are never compromised. Businesses and properties that are detrimental to the people and the environment are prohibited, no matter their owners' willingness to pay revolutionary taxes." However, in January of 2011, Jorge "Ka Oris" Madlos stated that although the NPA would like to drive mining companies away, it is currently unable to do so. Consequently, "they better pay taxes" (Arguillas 2011b). In eastern Mindanao, the NPA collects from one to three percent of all gross mining revenue in the form of revolutionary taxes and according to Madlos, "it is our inherent right to tax the businessmen in our area and even if you do not call us a government, as a revolutionary movement that is our inherent right" (Arguillas 2011b).

What makes revolutionary taxes so problematic is the concern that this may perpetuate the conflict between the NPA and the AFP while stifling forms of economic activity other than mining (Holden and Jacobson 2007b). With the financial resources provided by the revolutionary taxation of mining companies, the NPA can purchase new and better weapons thus enhancing their capacity to wage war and increase the level of violence in society. Meanwhile, other economic actors that are unable to pay revolutionary taxes are precluded from conducting business and become crowded out of the economy. This tendency locks the economy of the archipelago into an economy even more dependent on mining and acts as what Ross (1999, 321) calls a "violent version of the resource curse."

Mining projects as a source of weapons for the NPA

Mining projects also prolong the conflict between the NPA and the AFP by acting as a source of weapons for the NPA. With a few notable and important exceptions,[2] the NPA has been had to be self-sufficient in acquiring its own weapons. The principal way weapons have been acquired is through what are called *agaw armas* (gun-grab) raids where the NPA attacks an outnumbered AFP unit and takes its weapons.[3] Mines make excellent sites for *agaw armas* raids (as demonstrated in Table 6.1) for two reasons: first, mines always have armed security guards to protect them from the criminal element of Philippine society; second, mines are stationary targets and this allows the NPA "to conduct a thorough reconnaissance and to pick the most opportune time to strike" (Corpus 1989, 96). With mining projects being located in areas with substantial NPA activity, the NPA is given another source of weapons and this further entrenches the conflict into the fabric of the archipelago's society.

Mining as an NPA grievance mechanism

The threats posed to the vulnerable livelihoods of the poor and marginalized by large-scale mining can also exacerbate the conflict between the NPA and the AFP by acting as a grievance mechanism driving people into the NPA. Grievances can be thought of as social issues that insurgents can blame for life's

---

2 On two occasions during the 1970s, the NPA unsuccessfully attempted to import arms from the People's Republic of China. During 1981, 200 Automat Kalashnikovs were imported after being acquired from another national liberation group and in 1987, an unsuccessful attempt was made to import arms from North Korea (Caouette 2004).

3 Major Onting Alon, the civic affairs officer of the Philippine Army's Sixth Infantry Division, described an NPA *agaw armas* raid he experienced while on patrol as a lieutenant on 1 May 1990 in Davao Oriental. At least 200 NPA fighters ambushed his force of 60 soldiers. Major Alon's best friend, a fellow lieutenant, was killed, and the AFP lost their M60 machine guns to the NPA. Major Alon described it as the worst day of his life (Alon 2005).

problems; cadres can assess grievances and then articulate their insurgency as a solution to these issues (United States Army and United States Marine Corps 2007). Given the problems associated with large-scale mining, the NPA has long been critical of mining and attributed much of the *kahirapan* (difficulty) of life encountered by the poor to it. As Jose Maria Sison, writing under the *nom de guerre* Amado Guerrero ([1970] 2006, 95), stated: "Mines involve the direct seizure of land from the peasants and national minorities and also the destruction of wide expanses of agricultural fields as a result of the flow of mineral and chemical wastes in rivers." Forty-one years later, an article in *Ang Bayan* (2011, 1) reiterated this same theme stating:

Mining operations cause incomprehensible destruction: they trample on the nation's sovereignty. They cause widespread land grabbing. They plunder the nation's natural resources and destroy the environment. They poison our waters, crops, fishing grounds, pasture lands and other natural sources of livelihood.

In Chapter One the rebellion on the island of Bohol from 1744 to 1829 was discussed. In writing about this rebellion, the famous Filipino historian Renato Constantino (1975, 102) stated, "Three thousand people would not have abandoned their homes so readily and chosen the uncertain and difficult life of rebels had they not felt themselves to be the victims of grave injustices and tyrannies." When a mining related environmental disruption deprives poor and marginalized people of their livelihoods and thrusts them from subsistence into destitution, as was the case with the residents of the island of Marinduque after the tailings spill at the Marcopper Mine (and as was the case with the residents of the Albay Gulf after the tailings spill at the Rapu-Rapu Polymetallic Project), these people are likely to regard themselves as the victims of what Constantino would call "grave injustices and tyrannies." This is precisely the type of grievance that can drive people into the ranks of the NPA. According to Caouette (2004, 696), a renowned expert on the NPA, "people engage in violent collective action because it makes 'sense' given everything else" and "there are many potential recruits, particularly in rural areas where the CPP has focused its effort" (Caouette 2004, 697). Mines are developed in the Philippines with minimal public participation and these mines can then abruptly transform the poor from subsistence into destitution. With no input into mine development and with everything to lose should something go wrong at a mine, violent collective action may well appear imminently sensible. In the words of Father Peter Geremia, the Tribal Filipino Program coordinator for the Diocese of Kidapawan on the island of Mindanao,

"If there was no mining, there would be one less reason for people to join the insurgency" (Geremia 2005, interview).

The militarization of mining areas

Many mines in the Philippines are located in areas with NPA activity (Figure 7.7) and the NPA has attacked these mines (Table 6.1). This has led to 92 percent of the 494 respondents to the 2010/2011 Fraser Institute survey of mining companies stating that they view the security situation in the Philippines as something deterring investment in the islands (Fraser Institute 2011). The government is determined not to allow the NPA to thwart its mining-based development paradigm and consequently, the AFP has militarized many areas in the vicinity of mining projects as a way of providing security for them (Holden et al. 2011). On 8 February 2008, after the 1 January 2008 NPA attack on the Tampakan Project, the Investment Defense Force (a special AFP command) was created "to protect vital mining infrastructures and projects from those who stand in the way of development" (Ilagan 2009, 121).

What makes the militarization of mining areas troubling is the tripartite overlap between mining, indigenous peoples and NPA activity (Holden et al. 2011). Mines are located in mountainous areas (Figure 4.13), indigenous peoples are found in mountainous areas (Figure 4.16) and the NPA are located in mountainous areas. This confluence of geology, anthropology and insurgency has led to a substantial militarization of mining areas. This has, however, generated concern that the AFP may be militarizing areas near mining projects – which are populated by indigenous peoples – in order to intimidate their inhabitants into discontinuing their opposition to mining (Holden et al. 2011). When this happens, indigenous people opposed to mining are accused of rebellion or of engaging in terrorist activities. Just as the resource curse postulates a crowding out of other types of economic activity, the location of mining amid militarization crowds out competing nonhegemonic discourses, which are viewed as seditious and unworthy of being heard.

The province of Kalinga in the Cordillera has been heavily militarized by the following groups: the AFP; a PNP Regional Mobile Group; paramilitary Citizen Armed Forces Geographical Units (CAFGUs); and by the Cordillera People's Liberation Army (CPLA) – a paramilitary group staffed by Igorots (Holden et al. 2011). This is disconcerting because both CAFGUs and the CPLA have become notorious for their history of brutality. Ostensibly, this militarization has been encouraged to provide security for the Tabuk Copper Project (Figure 2.2 and Table 2.1), but people living in Kalinga have experienced heavy military harassment including frequent interrogations by the AFP. Although there is some

NPA activity in the area, there are concerns the government is exaggerating this as an excuse to crackdown on antimining activists.

On Mindanao, the AFP also militarizes areas to suppress the opposition of indigenous peoples to mining (Holden et al. 2011). In particular, the AFP has formed task force "Gantangan," a group of paramilitary forces staffed by Lumads similar to the CPLA, which spreads terror and fear among people who are suspected of being NPA sympathizers. The harassment and threatening of indigenous people involved in antimining activism becomes a matter of concern when one takes into account the spate of extrajudicial killings discussed in Chapter Five. Some in the islands even adopt a view that the militarization of mining areas inhabited by indigenous people is done to intimidate them into providing their consent under the Indigenous Peoples Rights Act. Father Romeo Catedral attests that there have been instances where indigenous leaders have consented to mining out of fear. For example, one indigenous leader consented to the Tampakan Project only after his brother had been killed by the AFP (Catedral 2005).

At TVI Pacific's Canatuan Gold Project, (Figure 2.2 and Table 2.1) controversy surrounding the impact of that Canadian mining company's paramilitary security force upon the indigenous Subanon living near the mine led to a human rights impact assessment being conducted by the International Centre for Human Rights and Democratic Development (2007), an institution created by the Canadian Parliament. At this mine, security is provided by a Special CAFGU Active Auxiliary (SCAA), a security force trained and equipped by the AFP but funded by the mining company. This human rights impact assessment concluded that the presence of the SCCA had not only caused the mine to fail to provide benefits to the Subanon, but it had also imposed costs on them by having "a negative impact on their ability to enjoy the human right to self-determination, to human security, to an adequate standard of living, to adequate housing, to work and to education" (International Centre for Human Rights and Democratic Development 2007, 38). All of these rights are guaranteed by a number of international human rights treaties ratified by the Philippines (International Centre for Human Rights and Democratic Development 2007). Similarly, at the Didipio Copper-Gold Project (Figure 2.2 and Table 2.1), operated by Australia's Oceana Gold, the Commission on Human Rights of the Philippines (CHR) (2011) held that the establishment of armed checkpoints in Barangay Didipio resulted in the unjust restriction of the resident's social and economic activities and this constituted a violation of their right to security of the person.

In addition to concerns about the NPA attacking mining projects, another reason why the AFP militarizes mining areas inhabited by indigenous people is a view that indigenous people are ripe for recruitment by the NPA. Major

Randolph Cabangbang, the spokesman for the Eastern Mindanao Command, stated that 70 percent of all NPA members in Eastern Mindanao are indigenous people and the NPA specifically tries to recruit indigenous people as part of their doctrine, particularly in mining areas (Cabangbang 2007). Many indigenous people are taken advantage of by the NPA because of their illiteracy and *datus* (Lumad leaders) encourage their people to join the NPA in an effort to reclaim their land (Cabangbang 2007).

The NPA has indicated that it recruits from all sectors of society so it cannot be denied that it recruits indigenous people (Holden et al. 2011). However, a heavily militarization of areas inhabited by indigenous people could actually lead to more indigenous membership in the NPA as people react to the presence of troops in their communities. This is what may be referred to as an increase in the "biographic availability" of the NPA (Caouette 2004, 229). In the 1970s in the Cordillera, when the government attempted to develop the Chico River hydroelectric project, the proposed hydroelectric dam was fiercely resisted by the Igorots and the vicinity of the project was heavily militarized (Finin 2008). Then, when Macliing Dulag (a highly respected Igorot leader) was killed by the AFP on 24 April 1980, the militarization led to even greater resistance by the Igorots who found the NPA an effective vehicle for resisting the dam (Holden et al. 2011). Similarly at the Mount Apo geothermal project on the island of Mindanao in the 1990s, the government developed the project on lands inhabited by Lumads who resisted the encroachment onto a mountain they viewed as sacred (Alejo 2000). To quell indigenous opposition to the project, the area was militarized and after the area was militarized, the NPA increased its presence. Many see today's militarization of mining areas (done because the indigenous inhabitants of such areas could be recruited into the NPA) as a self-defeating strategy (Holden et al. 2011). With mining areas being transformed into armed camps where the residents are subjected to martial law, it is difficult to see how mining brings benefits to current generations.

The illusion of mining's ancillary benefits

The penultimate problem with relying upon mining to act as an agent of economic development is the reliance upon mining projects to bring ancillary benefits to adjacent communities. Advocates of mining frequently like to emphasize the corollary benefits of mining projects such as improved roads, schools and health care clinics. As Mining, Minerals, and Sustainable Development (2002, 203) wrote:

There can be significant infrastructure improvements with the construction of a large mine. Most mining operations of any size are served by airstrips,

roads, water supplies, sanitation systems, and electricity. If these are restricted
to use by the company, and designed solely for company objectives, they
may be of little relevance to anyone else. With some advance planning and a
willingness to consult with the community, however, these can bring lasting
benefits at little or no added cost [and] the development of infrastructure
may facilitate other forms of economic activity such as tourism.[4]

The difficulty with relying on mining companies to provide such benefits
is the simple fact that, as Bishop Juan de Dios M. Pueblos (of the Diocese
of Butuan), stated, "Mining companies are not aid agencies" (De Dios
M. Pueblos 2005, interview). In the words of the Justice Peace and Integrity
of Creation Commission of the Association of Major Religious Superiors of
the Philippines (2009, 1):

> To look upon mining companies as agents of development is however
> problematic. Many problems are encountered when countries pursue
> development strategies through encouraging private investment. Mining
> companies are not, by nature, altruistic; they are in business to make a
> profit and if they do not make a profit, they do not stay in business for
> very long. In fact the private sector can address sustainable development
> concerns as long as adequate profits can be maintained. Mining companies
> exist to make profits not to help communities.

Cases on the ground also provide anecdotal evidence that mining's ancillary
benefits are illusory. At the Rapu-Rapu Polymetallic Project, the mining
company made a number of commitments but, in reality, very few of these
commitments were properly implemented (Oxfam Australia 2008). The
company promised to develop a piggery project to stimulate an alternative
livelihood to fishing; this project failed because the company stopped providing
loans before the mine became fully operational and people could not afford
to feed the piglets. When the piggery project was unsuccessful, a goat-rearing
project was planned but never implemented. When the mine was proposed,
the company promised to provide free electricity to the residents of Rapu-
Rapu Island; initially, this was only made available to selected individuals and
only made available to all residents of the island after the regular electricity
supply on the island was damaged by a typhoon during 2006. It was made
clear though, that once the mine closed, people would have to pay for their
electricity. This temporary provision of electricity would have the effect of

---

4 Given the visual impacts of mining discussed in Chapter Three, the authors find it
difficult to envisage how a postmining landscape will be conducive to tourism.

improving people's living conditions during the lifetime of the mine, only to then reduce their living conditions upon the cessation of mining operations. The company also promised to construct water tanks and connecting systems to supply water to people near the mine, but this water project was never completed because the company did not pay the contractors. "On the basis of these observations, it was clear that very few, if any, resources had been spent on effective and long-lasting community development" (Oxfam Australia 2008, 35). Stephen Davis was hired by Australia's Western Mining Corporation (WMC), the original project proponent of the Tampakan Project, to implement its indigenous people's policy. Davis even went so far as to admit that that mining companies are ill equipped to provide benefits to communities adjacent to their operations. As Davis (1997, 242) wrote, "WMC is a minerals company and, as such, does not have the specialized capacity to deliver community programs in all its mineral development projects."

If mines truly generate prosperity for the communities hosting them, why is it that mining communities are among the poorest in the Philippines? According to the World Bank (2010, 12), "Poverty incidence among households whose family heads were employed in mining and quarrying were almost as high as in agriculture." This means that poverty rates in mining communities often exceed the 40 percent level (Department of Environment and Natural Resources Climate Change Office 2010). The rural poverty rates in the islands were displayed on a province-by-province basis in Figure 1.3 and in Figure 4.15, mine locations were overlaid upon these rural poverty rates. The overlap between rural poverty and mining was displayed in Figure 4.15 to demonstrate the proximity of the vulnerable members of the rural poor to mining but it also demonstrates that mining fails to provide benefits for members of current generations. If mining is such a source of prosperity, why does one see such high poverty incidence where mines are located? The spatial correlation between rural poverty and mining demonstrates that in the Philippines, mining has the same economic effect as it does in the United States where, according to Power (2005, 96), "Mining communities are noted for high levels of unemployment, slow rates of growth of income and employment, high poverty rates and stagnant or declining populations."

The finite nature of mineral deposits

According to Bishop Juan de Dios M. Pueblos (2005), the benefits mining companies provide do not last forever as the minerals will eventually be depleted. Indeed, the finite nature of mineral deposits is the final and most intractable problem with pursuing a mining-based development paradigm. Mines have limited lifetimes and inevitably must close when the ore deposit is exhausted (Clements 1996; Lima and Suslick 2006; Power 2005, 2008). "Mines have finite lifetimes; the ore deposit is discovered, it is mined and then the mine is closed. The economic benefits of a mine only last throughout the lifetime of a mine" (Justice, Peace and Integrity of Creation Commission of the Association of Major Religious Superiors of the Philippines 2009, 1). "The stock of mineral resources is finite and the mine life is merely the length of time necessary to extract the ore body" (Lima and Suslick 2006, 88). In recent years, the uncertainty generated by volatile mineral prices has placed substantial pressure on mining companies to extract minerals as rapidly as possible (Power 2008). Acting in response to this pressure and using new technologies, modern mines often have lifetimes as short as between 8 to 15 years (Power 2008). The limited lifetime of an individual mine may not be a serious concern when discussing mining at the national level because as some mines become exhausted, new mines will be developed and the closure of older mines is made up for by the opening of newer ones. However, at the level of the individual community, mine closure can be catastrophic and mining communities often cease to exist when the deposit is mined out (Clements 1996). In mining towns, the economic life of the community is almost totally dependent on the company. As Roque and Garcia (1993, 112) wrote:

The company is sometimes the sole provider of schools, churches, clinics, and sometimes electrical power. In essence, the mining towns are *haciendas* with a different product. The managers of the company are descendants of the *hacendados*, and the workers are usually of peasant stock.

Mining is a skills-specific industry and the small number of people who become employed in mining enter the industry at a young age, learn and develop skills and end up spending most of their lives in the industry; losing such a job does not provide one with a skill set readily transferable to employment in a different trade (Bello et al. 2009). Mineworkers located at the mine with their families cannot easily relocate to another mine site and often have nowhere to go and no one to turn to when a mine closes down (Bello et al. 2009). Consider the Toledo Copper Project, in Toledo City, Cebu, which was closed by Atlas Consolidated Mining in 1995 (Parreno 2008). Although the recent surge in copper prices has led to a reopening of this mine, the effect upon the community of the 1995 closure was disastrous. As Banzon (2004, 38) wrote:

Thousands of Toledo breadwinners lost their only means of livelihood and most of the peripheral businesses folded up. The once busy village road was shrouded in darkness and the frustration and pain of the retrenched workers made the village a virtual ghost town. Today the residents are trying to pick up the pieces of the lives shattered with the cessation of Atlas Mining operations. Those who lost their jobs tried to find employment hoping that someday, God would come to the rescue and carry through the spot where the mining firm has left off.

Mining Imposes Costs on Future Generations: A Lack of Intergenerational Equity

Mines require perpetual attention upon closure

The difficulties discussed so far with relying upon mining to act as an agent of economic development (limited employment opportunities, limited tax revenues, a lack of linkages to other industries, volatile mineral prices exhibiting a long-term downward trend, the resource curse thesis, the exacerbation of the conflicts with the MILF and the NPA, limited ancillary benefits from mining and the finite nature of ore deposits) all indicate mining fails to provide benefits to present generations (the first component of sustainability in Table 7.1). However, just as mining fails to provide benefits to present generations, it also places expenses upon future generations (the second component of sustainability in Table 7.1), and thus fails to exhibit intergenerational equity. Mines, as made clear in Chapter Three, require perpetual attention once closed. In the United States, more than half of a group of 156 mines studied between 2002 and 2003 will require attention for periods of time ranging from 40 years to perpetuity (United States Environmental Protection Agency 2004). Consider, for example, the Berkeley Pit in Butte, Montana in the United States of America as discussed in Box 7.1.

The substantial expense of mine reclamation

Examples from the United States

In addition to requiring permanent attention, many mines also require substantial expenditures upon closure. In the American state of Colorado, the Summitville mine suspended operations in 1991 but not before it caused a 17 kilometer stretch of the Alamosa River to be declared biologically dead due to acid mine drainage and cyanide spills (Gedicks 2001; Young 2000). It has been estimated that clean-up costs at this mine will range from between USD 150 million to USD 170 million (Gedicks 2001; Young 2000). In the state of Montana, the conjoint Zortman and Landusky mines will cost between USD 50 million to USD 190 million to reclaim (Holden et al. 2007). In the United States, the 156 mines studied between 2002 and 2003 have the potential to cost between USD 7 billion to USD 24 billion to clean up (United States Environmental Protection Agency 2004).

Mine reclamation requirements in the Philippines

The Philippine government, aware of the potential costs inherent in mine reclamation, acted to initiate a set of mine reclamation rules in 2005 with the implementation of DENR AO 2005–06 and DENR AO 2005–07. The first of these administrative orders, DENR AO 2005–06, implemented a set of guidelines for a system of mandatory environmental insurance coverage requiring all mining projects to obtain an environmental performance bond to provide funds for ameliorating any environmental liabilities generated by the mine and to obtain environmental liability insurance (Bautista 2008). The mining industry vigorously resisted the provisions of DENR AO 2005–06 calling it a redundancy and a de facto tax upon the mining industry (Bautista 2008). Consequently, on 21 February 2006, DENR AO 2005–06 was suspended indefinitely. The second of these administrative orders (DENR AO 2005–07) has provisions requiring project proponents to contribute to a mine monitoring trust fund (MMTF), so as to provide a source of funds to pay for mine monitoring (Bravante and Holden 2009). There are provisions requiring the deposit of funds to facilitate a mine rehabilitation fund, so funds can be available to pay for the rehabilitation of a completed mine. Proponents are also required to develop a final mine rehabilitation/decommissioning plan (FMR/DP) (Bravante and Holden 2009). There are also provisions in the FMR/DP requiring the setting aside of up to PHP 5 million (approximately USD 115,000) every year during the operation of the mine so as to ensure that when the mine is closed, there will be funds available for its remediation (Bravante and Holden 2009).While these requirements have a laudable motive behind them, it can be argued that they do not go nearly far enough.

An inadequate amount of funds for mine reclamation

The first serious problem with these provisions is the amount of money that mining project proponents are required to set aside. The MMTF must be no less than PHP 150,000 (approximately USD 3,500) and the mine rehabilitation fund is fixed at a maximum annual contribution of PHP 5 million (approximately USD 115,000) (Bravante and Holden 2009). As of February 2010, the average amount of money in approved final mine rehabilitation/decommissioning plans was less than USD 500,000 per mining company (Department of Environment and Natural Resources Climate Change Office 2010).

A finite time period for the management of a perpetual problem

The second major problem is that mine reclamation plans only provide for proponents to be responsible for their projects for a ten-year period after closure. In view of the fact that processes such as acid mine drainage involved in mine reclamation operate on a geologic time scale, ten years is an exceedingly short time period. Also, mine reclamation plans are to be developed using risk-based methodologies that estimate the probability of an event (such as a typhoon or earthquake) adversely impacting the mine after it has been closed. Mine reclamation plans are not required to be subjected to the higher standard of protection emanating from using a worst-case scenario, which would look at what would happen when a typhoon or earthquake adversely impacts the mine. With the high vulnerability of the Philippines to natural hazards, the latter approach would be more prudent as this would generate more conservative design estimates and minimize the potential for a catastrophic accident. It is important to use worst-case scenarios in designing closure plans instead of using risk-based methodologies, wherein the risks of accidents are estimated using probabilistic methodologies. Assessing risk is a highly subjective process and risk-based methodologies do not provide the higher standard of protection emanating from using a worst-case scenario. This was the point made clear by Shrader-Frechette (1993) in the discussion of deep geologic disposal of spent nuclear fuel. Since nuclear waste will remain radioactive for millions of years, events such as earthquakes or volcanic eruptions cease being probabilities and become certainties. In such a context, it is not acceptable to discuss what will happen if an earthquake impacts the nuclear waste repository because over millions of years, an earthquake will impact the waste repository so precautions for such an event must be taken into account when it is designed. Given that processes, such as acid mine drainage, also operate on a geologic time scale, the analogy of a spent nuclear fuel repository is imminently reasonable.

The lack of a public fund for mine reclamation

In addition to the problems with the short time horizon of these mine reclamation plans and their reliance upon risk-based methodologies, the provisions in them requiring the annual setting aside of up to PHP 5 million (approximately USD 115,000) a year during the life of the mine so as to ensure that when the mine is closed there will be funds on hand for its remediation are also inadequate. Should the mining project proponent become insolvent, cease operations and cease setting aside money, this will generate an unfunded environmental liability. There have been many situations where mining companies have gone bankrupt during the course of the mine's operations and left behind a mine requiring a publically funded reclamation. Perhaps the government should require the proponent to post a bond providing the funds reasonably foreseeable for mine reclamation at the beginning of the project. This way, if premature closure occurs, the public has the money on hand to fund mine reclamation.

Also, the Philippines has nothing akin to the American Superfund law[5] to ameliorate unfunded environmental liabilities although the Alternative Mine Bill (AMB) would create such a fund with its requirement that all mineral agreement holders deposit PHP 5 million (approximately USD 115,000) a year into a fund maintained by the government to be used for redressing the effects of calamities relating to mining activities. In the absence of a Superfund, the state will have to adopt the costs of reclaiming an abandoned mine. The inadequacies of programs for the remediation of unfunded environmental liabilities are a problematic aspect of implementing a mining-based development paradigm in the Philippines. As Bravante and Holden (2009, 539) wrote:

> The archipelago has potential to become dotted with numerous large-scale mines. These mines operate without being subjected to any bonding requirements. Should one of these mines be abandoned, there will be inadequate programs in place to clean up the mine. Funding for mine reclamation will, presumably, then have to be taken from general tax revenue.

In their zeal to attract mining investment, Philippine policymakers have allowed this situation to develop because they do not want any potential deterrent to mining investment coming from requiring project proponents to post reclamation bonds or to contribute to something akin to the Superfund. This has led to situations (such as that at the Marcopper Mine on the island of Marinduque) where mining related accidents have generated calamities that have devastated local communities and the members of those communities have been left uncompensated. Residents of the island of Marinduque have never received compensation for the harm inflicted on them and are currently in the process of attempting to sue the remnants of the Marcopper Mining Corporation for compensation.[6]

---

5 In the United States, the Comprehensive Environmental Response, Compensation, and Liability Act 26 USC 4611-4682 created a pool of funds (referred to as the "Superfund") for the rectification of unfunded environmental liabilities.

6 *Rita Natal et al. v. Marcopper Mining Corp.* (Regional Trial Court Civil Case No. 01-10) was filed in 2001 and is making its way through the courts. Residents of Marinduque filed a mandamus application with the Supreme Court of the Philippines in 2010 seeking a court order compelling the trial judge to expedite the proceedings.

The sacrifice of provisions requiring mine rehabilitation on the altar of attracting mining investment is an excellent example of the mentality prevalent in the developing world that "a bill deferred seems almost as good as a bill unpaid" (Westin 1992, 187). If one can keep putting off, and putting off and putting off funds for mine rehabilitation, it is almost as if such funds never need be allocated. However, the irony of this is that it will prove self-defeating because it will lead to a series of spectacular disasters where no one receives compensation. In the meantime, though, the investment dollars keep coming, the mining industry stays happy and the wisdom of the prophets at the World Bank remains adhered to.

The foregoing problems associated with the (un)sustainability of mining caused Westin (1992, 202) to declare mining a "doomed activity" and to state that "eventually many countries will deplete their stocks of mineral wealth without establishing alternative industries or agricultural, leaving large impoverished populations with nowhere to turn for their livelihoods." Indeed, Westin (1992, 202–3) went so far as to draw an analogy between mining and drug abuse:

> For many developing countries, mining activities are perilously similar to addictive drugs that ultimately render the user impoverished and exhausted. One difference is that with addictive drugs, the drama plays out in one lifetime; with imprudent mining practices, the drama spans several generations, with inconsistent amounts of pleasure or suffering among the generations.

A Lack of Widespread Benefits: A Lack of Intragenerational Equity

Large-scale mining fails to comply with both of the requirements for sustainability stipulated in Table 7.1. Given these two failings, it can be said that a mining-based development paradigm fails to comply with the necessary condition for sustainability. Large-scale mining conducted by corporations is also an activity that does not result in an equitable sharing of benefits among members of the current generation and thus fails to comply with the necessary condition for development in Table 7.1. The problems inherent in funding the costs associated with the reclamation of the Zortman and Landusky mines in the state of Montana provide an excellent example of how mining fails to be sustainable and Box 7.1 discusses how the lack of widespread benefits accruing from mining in Montana provides an excellent example of how mining unsuccessfully provides development.

As discussed in Chapter One, the Philippines are dominated by a powerful oligarchy. With a powerful elite dominating policymaking in the archipelago,

Box 7.1. Mining in Montana: Long-term costs and a lack of widespread benefits

Despite the vast spatial and social distances separating them, the American state of Montana and the Philippines have a profound historical connection. In 1898 the 1st Montana Infantry Regiment was deployed to the archipelago as part of the United States Army's 8th Corps under the command of General Wesley Merritt (Hines 2002). The 1st Montana was one of the first American units to engage the Filipinos on the outskirts of Manila on 4 February 1899, and served with such distinction in the islands that in 1905 the state legislature adopted its regimental flag as the state flag (Hines 2002).

Montana and the Philippines also have a connection in that both places have a long history of mining. Mining played an essential role in the development of Montana; the historical importance of hardrock mining in Montana can be readily demonstrated by reference to the state's nickname, "The Treasure State," and the state motto, "Oro y Plata" (Spanish for gold and silver). Mining began in Montana with gold mining in the 1860s, and was followed by silver mining in the 1870s and copper mining in the 1880s (Holden et al. 2007). By the 1890s the dominant mineral being extracted from Montana was copper, and the dominant mining company was the Anaconda Copper Mining Company, the world's largest supplier of nonferrous metals (Swibold 2006). Originally the copper mining carried out in the 1880s was done by using underground mining but, by the 1950s, competitive pressures from lower cost open-pit mines forced the Anaconda Copper Mining Company to switch to open-pit mining, and an immense open-pit mine called the Berkeley Pit (Figure 7.8) was developed, which destroyed a large portion of the town of Butte (Dobb 2002).

By 1982, after operating for 27 years, mining was discontinued and the Berkeley pit began flooding with water so acidic it could "liquefy a motorboat's steel propeller" (Dobb 2002, 312). "Water will always migrate into the pit from the highly fractured and heavily mined bedrock that surrounds it; the infernal receptacle will always be cursed. Not for a hundred years, not for a thousand, but always" (Dobb 2002, 328). While the Berkeley Pit provided jobs for a generation of workers in Butte, its long-term effects will last forever. As Dobb (2002, 331) wrote:

The pit was excavated in one generation, the dozens of underground mines that surround it in five or so, but the aftereffects of these

Figure 7.8. The Berkeley Pit in Butte, Montana

Photo credit: Montana Historical Society Archives.

engineering feats will be felt for hundreds of generations, until the next ice age or geologic cataclysm, a perpetual problem in need of a perpetual solution.

Anaconda Copper was such a powerful force in the state of Montana that it was referred to as simply "the Company" and it ruled Montana from its New York City headquarters with an iron fist from the 1890s until the 1970s (Holden et al. 2007). According to Malone and Roeder (1976, 323), "the Anaconda Company wielded such enormous power in Montana that the state gained the unenviable reputation of being nothing more than a corporate asset." Dobb (2002, 317) stated, "Anaconda maintained a more comprehensive hold on Montana's natural resources, government and people than any other corporation in any other state", while Swibold (2006, xiv) wrote "the company loomed over Montana's political and economic consciousness." Anaconda owned most of Montana's mines, hotels and logging operations (Holden et al. 2007). Anaconda also owned most of the major newspapers in Montana, which wrote scathing editorials criticizing anyone, and indeed everyone, who dared oppose it (Swibold 2006). This corporation would do all that it could to thwart any attempts to redistribute wealth away from it to either the state of Montana or to its own workers. In 1903, unable to acquire a favorable decision from judges in Butte regarding the ownership of some ore deposits, Anaconda shut down all of its operations statewide and threw 6,500 workers out of work (Dobb 2002). In 1919 Louis Levine, an economics professor at the University of Montana, called for the imposition of higher taxes on the mining industry in an attempt to make Anaconda responsible for a greater contribution to the state's finances (Dobb 2002). Anaconda was so outraged by this it brought pressure on the University of Montana and had Levine fired (Holden et al. 2007). The company consistently resisted any unionization of its workforce, engaging in mass firings of suspected socialists, and often used convicts (supplied by the warden of the state penitentiary) as strikebreakers (Swibold 2006). Eventually, the nationalization of Anaconda's extensive, and well-endowed, Chilean mines (begun in the 1960s under President Eduardo Frei and completed in 1971 under Salvador Allende) crippled the company and freed the people of Montana from its dominion (Holden et al. 2007). Nevertheless, during Anaconda's domination of the Treasure State over USD 25 billion worth of mineral resource wealth was extracted, and was concentrated in the hands of this corporation and removed from the state (Dobb 2002).

What is most problematic about the lack of widespread benefits emanating from mining in Montana is the fact that this was able to occur notwithstanding the long tradition of strong democracy prevailing there (Holden et al. 2007; Malone and Roeder 1976; Swibold 2006). Montana is a place that has implemented "an impressive array of progressive reforms designed to popularize democracy," including voter initiatives giving citizens "the power to make or reject laws in the face of legislatures controlled by corporate interests" (Swibold 2006, 117–18). If large-scale corporate mining can result in a high concentration of benefits in Montana, where there has been a long tradition of strong democracy, what will happen with large-scale corporate mining in the Philippines, a country dominated by a powerful oligarchy?

it is doubtful whether there will be a widespread distribution of mining's benefits among all sectors of society. "The state [has been] subjugated by a succession of ruling elite factions to serve narrow interests instead of the larger goals of sustainable development and social justice" (Bello et al. 2009, 244). The islands have "a society where economic progress hardly trickles down" (Kirk 2005, 123). In the archipelago, the poor do not receive any lasting benefits from the "ups" of the economic cycle but they always slide to lower levels of poverty with its "downs" (Roque and Garcia, 1993). Carroll (1986, 31) expanded upon this, writing "The national income, in seeming contradiction to the laws of gravity, has tended to 'trickle upward' to the highest reaches of the social system."

Many members of the oligarchy own – either in whole or in conjunction with foreign investors – the mining companies operating in the archipelago (Bello et al. 2009; Tujan and Guzman 2002). The Benguet Corporation is owned partly by the Ayalas and the Sorianos; Lepanto Consolidated Mining is owned by the Yaps, Sycips and Velayos; while Hinatuan Mining, Rio Tuba Mining and Taganito Mining are owned by the Zamoras (Bello et al. 2009; Tujan and Guzman 2002). With the oligarchy controlling the operation of the state and concomitantly operating large portions of the mining industry, it is highly unlikely that there will be any widespread distribution of mining's benefits among all sectors of society. To Rovik Obanil, the Philippines are an example of booty capitalism; "it is really the rich people getting richer and very little trickles down to the communities" (Obanil 2009, interview). Ricardio Saturay views mining is an example of a plunder economy in that the mining industry is controlled by elite families (Saturay 2009). To Jesus Garganera, in the Philippines, destructive activities such as mining are taking place because the rich and powerful benefit from them (Garganera 2009). Should the government continue with its aggressive promotion of large-scale corporate mining, there will be "a concentration of benefits among a small number of beneficiaries and a democratization of mining's environmental costs among the poor engaging in subsistence activities" (Bravante and Holden 2009, 543). Such a concentration of benefits and democratization of costs is not development.

As indicated in Chapter Six, the rich and powerful members of society in the archipelago could actually benefit from the poor and marginalized being adversely affected by the environmental effects of mining. Not only does mining result in an unequal distribution of benefits among members of the current generation, it could, given significant instances of environmental degradation such as what occurred at the Rapu-Rapu Mine after the October 2005 tailings spills, result in a situation where the rich not only receive the majority of its benefits but also become richer as a result of its impact upon the poor. If the poor are displaced from their subsistence farming and fishing, they will enter the labor market and push down wages thus benefiting the rich and powerful by creating lower cost labor. This exemplifies how environmental change is not simply confined to changes in the quality of the biophysical environment but can also constitute altered power relations in society. Environmental change can result in some people being given an enhanced ability to control others while concomitantly reducing the ability of some people to resist being controlled by others. Bearing this thought in mind, something that not only results in a concentration of benefits and democratization of costs and also results in an enhancement of social inequality looks even less like development.

Mining as Development Aggression

The concept of development aggression

At the beginning of this chapter, the question of whether a mining-based development paradigm is an appropriate approach to be followed in the Philippines was posed. The interactions of mining's environmental effects with the natural hazards present in the Philippines will cause mining to disrupt the most important resource for those engaged in subsistence activities: a clean and healthy environment. It is unlikely that mining will act as an engine of economic growth, which generates so much prosperity that whatever instances of environmental disruption may occur can easily be compensated for by the "rising tide of prosperity" that "lifts all boats." In addressing the efficacy of this development strategy, the concept of sustainable development was used as a metric. Large-scale mining fails to provide benefits to current generations, it imposes costs on future generations and it does not generate an equitable sharing of benefits among members of the current generation; mining does not comply with either the necessary conditions for sustainability or the necessary condition for development. Given the potential for mining to intersect with natural hazards and generate disasters and given the failure of mining to comply with the metric of sustainable development, one may find the discussion provided by Bankoff and Hilhorst (2009, 691) of how mining can generate "development induced disasters" or "development aggression" useful.

The term "development aggression" is a term frequently used by members of Philippine civil society critical of the government's efforts to promote mining (Figure 7.9). Nadeau (2005, 334–5) defines development aggression as, "the process of displacing people from their land and homes to make way for development schemes that are being imposed from above without consent or public debate." Development aggression is also referred to by Nadeau (2002, 103) as "inappropriate development," which is "a globalizing economic and

Figure 7.9. Antimining protest, General Santos City, May 2005

political process coming from outside that severely damages a community's culture, social organization and environment." The International Coordinating Secretariat of the Permanent Peoples' Tribunal and IBON Books (2007, 186) defines development aggression as "development projects that destroy [a community's] traditional economy, community structure, and cultural values." In a discussion of mining and indigenous peoples Doyle (2009, 56) defined development aggression as the situation "whereby predominantly foreign companies, with assistance from the Philippine elite, have encroached upon their lands in their quest for natural resources, impoverishing rather than benefiting indigenous peoples." In Bishop De Dios M. Pueblo's opinion, development aggression is when someone else imposes a project on the people living in an area that makes them poorer (De Dios M. Pueblos 2005). Father Ramoncito Segubiense views mining as development regression because development should be an advancement of the community and mining does not lead to an advancement of the community, rather, its benefits only go to the owners of the mining company (Segubiense 2009). To Father Ramoncito, development should be something that makes the local community better off not worse off (Segubiense 2009).

The concept of appropriate development

In contrast to development aggression, appropriate development refers to a process leading to the qualitative improvement of all members of society (Nadeau 2002). The Commission on Human Rights of the Philippines (2011, 1) used a similar definition of development, in its investigation into the Didipio project being carried out by Australia's Oceana Gold stating, "The ultimate goal of economic development is to raise the quality of life of all people."

What Nadeau (2002) calls appropriate development, Broad (1988) calls genuine development and this is brought about through complying with three important principles: diversification of economic activities, programs centered on people and public participation. A diversification of economic activity leads to a reduction of dependence on the global economy, which prevents events in the developed world from disrupting life in the developing world. The volatility of mineral prices is an excellent example of the importance of diversification. Mineral prices can rise and crowd out economic activity in the Philippines, or they can fall and lead to mine closures that devastate local communities dependent upon them. Population centric development activities are essential because people should be the essential focus of all development programs. As Broad (1988, 233) wrote: "People – their dignity, their participation and their empowerment, as well as the satisfaction of their basic needs – should be the primary goal of any development effort." This entails "placing redistribution and equity policies at the very top of the adjustment priority list" (Broad 1988, 233). Given the tendency of mining to produce concentrated benefits, and widespread costs, this appears unlikely to happen under a mining-based development paradigm. Finally, there must be a participatory development process wherein people participate in making decisions that affect their lives. Inhabitants of an area must be able to decide what kinds of projects they want and what kinds of projects they can afford (Broad 1988). As Broad and Cavanagh (1993, 139) wrote, "The struggle for the environment and for control of resources requires a far more participatory notion of development." The weak provisions for public participation in the Philippine EIA system demonstrate how mining in the Philippines is largely an activity occurring without meaningful public participation and, therefore, is not an example of genuine development. In Chapter Four, parallels were drawn between mining and the Bataan Nuclear Power Plant (BNPP) in terms of how they are both high risk activities located in hazard-prone areas. The BNPP also suffered from a lack of meaningful public participation. Broad and Cavanagh (1993, 127) could find no Filipino who had anything positive to say about it; officials of the provincial government were not consulted about its construction and fisher-folk in the nearest town could only "swear at it and recall how tens of thousands of citizens protested against its construction."

Answering the Questionable Development Strategy

This chapter has shown that a mining-based development paradigm is not an appropriate approach to be followed in the Philippines. Mining-related environmental disruptions brought on by the interactions of mining's environmental effects and the natural hazards present in the Philippines will only serve to disrupt the ecology of the poor and end up impoverishing vulnerable communities adjacent to mining operations. Given the failure of mining to meet the needs of present generations, the costs imposed by mining upon future generations and the inequitable sharing of mining's benefits among members of the current generation, mining will not act as an engine of economic growth generating so much prosperity that whatever instances of environmental disruption may occur will easily be compensated for by the subsequent rising tide of prosperity that "lifts all boats." Attention now turns toward addressing some alternatives to large-scale mining. Are there alternatives to mining capable of ameliorating the poverty encountered by so many Filipinos? Or, alternatively, is this even an appropriate question to ask given the costs imposed by mining and the harm befalling vulnerable communities adjacent to mining projects?

Chapter Eight

IS ANOTHER WORLD POSSIBLE?

Mining: A Flawed Development Model

The preceding chapter demonstrated that a mining-based development paradigm is inappropriate in the Philippines. Mining-related environmental disruptions will disrupt the ecology of the poor and end up impoverishing vulnerable communities adjacent to mining operations. The instances of environmental disruption occurring when large-scale mining is located amid the natural hazards present in the archipelago will not be compensated for by a rising tide of prosperity lifting all boats. Opposition to mining is so pronounced in the islands that nine provincial governments (Figure 6.3) have passed moratoriums banning large-scale mining within their jurisdiction. These provincial governments are so concerned about the environmental effects of large-scale mining that they have gone so far as to ban it completely. To the residents of these provinces, a complete and utter absence of large-scale mining is preferable to any form of it. This is not a discussion of how mining can be implemented differently so as to better propel the residents of these provinces towards some teleological concept of development, rather this is a discussion of the residents of these provinces being happy living the way they currently are and not wanting mining to disrupt their sources of income.

Consider the residents of the province of Sorsogon, deprived of their livelihoods by the cyanide spill at the Rapu-Rapu Polymetallic Project. To these people, mining did nothing but plunge them into destitution; they would have been much better off had the mining company never turned a shovel on Rapu-Rapu Island. This is, again, the essence of the concept of development aggression: the situation where a development project is imposed on people that make them poorer.

When proponents of mining such as the Chamber of Mines of the Philippines, the Philippine government and the World Bank encounter such arguments, they will immediately respond by posing the question, "What is your alternative?" Opponents of mining are under no obligation to answer this question. Mining's proponents have failed to make their case that mining will act as a vehicle capable of accelerating the development of the archipelago; this does not make it incumbent upon the opponents of mining to prove that there is a viable alternative. The residents of Barangay Macambol have no responsibility placed upon them to demonstrate a viable alternative to mining. They are content living their lives as subsistence fisher-folk and, unless and until large-scale mining's advocates can unequivocally demonstrate that mining will not prejudice them, the residents of Barangay Macambol should be allowed to continue fishing and not be required to prove that there is a viable alternative to mining.

In many ways, what makes the adoption of a mining-based development paradigm problematic is the entire concept of development underlying it. Advocates of mining draw attention to the prolific mining industries established earlier in the histories of developed countries such as Australia, Canada and the United States (Power 2002). These advocates of mining maintain that only if the Philippines can establish a prolific mining industry, resembling those seen in nineteenth-century Australia, Canada or the United States, will the archipelago "catch up" with those countries and become a developed country. This concept of "catching-up development" is controversial as it is "based on an evolutionary linear understanding of history" (Nadeau 2008, 101). As Nadeau (2008, 101) wrote:

> The illusion of catching-up development is based on the false belief that the only appropriate model of an affluent society is that prevailing in the United States, Western Europe, Hong Kong, Taiwan, Japan and the Republic of South Korea. It implies that poor countries that follow the same path to industrialization and capital accumulation taken by modern industrial societies can achieve the same level of development. Yet, most of the richest nations of the world still have not attained a satisfactory level of development. In the United States, for example, real incomes and quality of life for many citizens have gone down steadily, especially in the post-9/11 period. Poor neighborhoods in Los Angeles and New York City, for example, are often dangerous places overtaken by drugs, gangs and violence. Also, there are many citizens living in deprived circumstances in some of the most highly developed countries of the world while the material conditions for most citizens in poor countries are getting increasingly worse.[1]

---

1 Nadeau's comments about declining real incomes and quality of life for citizens of the United States of America are appropriate. The United States is a country that routinely, and unabashedly, describes itself as the greatest country in the world. However, in 2011 the United States had a higher infant mortality rate (6.06 deaths for every 1,000 live births) than Cuba (4.9 deaths for every 1,000 live births) (Central Intelligence Agency 2011).

Instead of following the "catching-up development" paradigm, many members of Philippine civil society are calling for two things with a substantial scope for ameliorating the poverty encountered by so many Filipinos: land reform and the sustainable use of natural resources to meet local needs (Nadeau 2008).

Land Reform

The importance of land reform in the Philippines

Land reform, often referred to as "agrarian reform," is defined by Putzel (1992, xx) as "programs, usually introduced by the state, that have the intention of redistributing agricultural land to its tillers and providing them with secure property rights and the means to earn an adequate living." If development is considered to be the improvement of the material wellbeing of all people in society, then it should be focused on the sectors of the economy where most of the population lives and works (Putzel 1992). In the Philippines this is the agricultural sector (Figure 8.1), which is responsible for 40 percent of all employment and where there is widespread landlessness, near landlessness and inequality (Borras 2007).

Since many farmers in the islands have no secure access to land, they "can never be certain of meeting their basic needs for survival" (Putzel 1992, 22). "Agrarian reform," wrote PESANTEch (2000, 127),[2] can lead to "strong rural

Figure 8.1. Land reform, one of the most pressing needs in Philippine society

Photo credit: Keith Bacongco.

agricultural communities composed of farmers whose industry could – if properly encouraged and supported – generate surplus income that would make for a more dynamic rural economy… Agrarian reform [is] a necessary and potent strategy for eliminating poverty at all levels" and poverty is, without a doubt, "the antithesis of development."

Land reform and insurgency

In Chapter One the conflict between the AFP and the NPA was outlined and in Chapter Seven the problems this conflict poses when juxtaposed with large-scale mining were discussed. The lack of land reform in the Philippines is widely regarded as being perhaps the single most important factor contributing to the persistence of the NPA. This persists long after Fukuyama (1989), in his seminal piece of neoliberal triumphalism, declared history to be over and Maoism to be an anachronism.

Victor Corpus was a lieutenant in the Philippine Army and an instructor at the Philippine Military Academy who defected to the NPA in 1970 after leading an NPA raid on the armory at the academy. From 1970 to 1976, Corpus was a member of the NPA until surrendering in 1976. Corpus returned to the AFP in 1987 and wrote the book *Silent War.* According to him (1989, 109), "Only honest-to-goodness land reform can prevent desperate and starving people from rushing into the arms of the rebels." Of all of the problems faced by the poor and marginalized, the issue of agrarian reform is "the most urgent and primary" (1989, 183). To Corpus, the insurgency of the NPA is like cogon grass (a weed species that is difficult to eradicate) and its taproot "is the unequal distribution of land ownership in the country" (Corpus 1989, 183). As Corpus (1989, 185) wrote: "When I was still with the [NPA], I remember that the foremost issue that we used to exploit in arousing and mobilizing the masses to rise up in arms against the government centered on the agrarian issue."

Not one of the political administrations during the past one hundred years has seriously addressed the underlying cause of rural unrest posed by landlessness (Borras 2007). Failure to implement meaningful land reform means that all attempts to address insurgency are nothing more than attempts at "crushing the manifestations of social unrest without addressing their root causes;" all this does is extend the "Sisyphean cycle of repression and revolt" (McCoy 2009, 450).[3]

---

2 "PESANTEch" stands for "Paralegal Education Skills Advancement and Networking Technology Project," a project to assist the peasant sector in the Philippines being implemented by a group of progressive lawyers.
3 In Greek mythology, Sisyphus was condemned by the gods to roll a boulder up a hill, only to have it roll back down just before reaching the top of the hill, and to repeat this throughout eternity. Accordingly, a pointless or interminable task can be described as a "Sisyphean task."

The history of land reform in the Philippines

Land reform during the Commonwealth period: 1935–41

By the 1930s Central Luzon had become "a social volcano, constantly rumbling with discontent, erupting periodically in local revolts" (Karnow 1989, 272). To forestall the growing peasant unrest, the Commonwealth government created the National Land Settlement Administration (NLSA) in 1939 to organize resettlement of landless peasants from Central Luzon and the Visayas to public lands on Mindanao and in northeastern Luzon (Putzel 1992). It opened the Koronadal Valley and the Allah Valley in Cotabato and the Mallig Plains in Isabela but, by 1950, it had resettled only 8,300 families.

Land reform during the 1950s

In 1950, alarmed at the success of the Hukbong Mapagpalaya ng Bayan (Army of National Liberation or HMB) uprising, the United States sent Daniel Bell (the Undersecretary of the Treasury) to conduct a survey of economic conditions in the archipelago (Putzel 1992). "The Bell Misson's report was the first to propose widespread land redistribution as the only solution to agrarian unrest" (Putzel 1992, 85). Then, in 1951, the United States Mutual Security Agency commissioned Robert S. Hardie to study the tenancy problem in the islands and Hardie's report called for widespread land reform. As Putzel (1992, 85) wrote:

> His recommendations called for abolishing absentee ownership; low ceilings on land retained by landlords; government purchase of land and a ban on negotiations between landlord and tenant; a land price fixed by the government and compensation in non-negotiable government bonds; a flexible system of amortization payments by tenants; and tenant participation in the reform through land commissions.

Yet, at the same time that the Hardie report was released, Defense Secretary Ramon Magsaysay, under the tutelage of Lieutenant Colonel Edward Lansdale of the Central Intelligence Agency (the "ugly American"), began having success against the HMB through a combination of counterinsurgency tactics and the exploitation of the poor organization of the HMB (Caouette 2004). One seemingly progressive thing Magsaysay did was to organize the Economic Development Corps to resettle landless members of the HMB on public lands on Mindanao (Putzel 1992). While this program resettled a miniscule number of actual HMB members, it was a massive propaganda success that led to the voluntary migration of thousands of landless people to Mindanao – the "land of promise" – and it gave many people the impression that Magsaysay defeated the HMB by engaging in land reform. As Galula (1964, 13) wrote:

> Land reform looked like a promising cause to the Hukbalahaps after the defeat of Japan and the accession of the Philippines to independence; but when the government offered land to the Huks' actual and potential supporters, the insurgents lost their cause and the game.

This is, however, a highly inaccurate view. "Resettlement was not a solution because eventually the same tenancy practices would beset new areas of settlement" (Putzel 1992, 87). The migration of landless people from Luzon and the Visayas to Mindanao and their resettlement on public lands did not alter pre-existing distributions of wealth and power in society, hence it did not "constitute and promote redistributive reform" (Borras 2007, 27). "To settle people on new land and to develop it for agricultural use does not involve any basic alteration of the property rights of existing landowners" (Borras 2007, 27). "This critical failing left the underlying social inequality (the root cause of the country's endemic violence) unchanged, and meant that much of his political legacy was one of transitory, essentially palliative political gestures" (McCoy 2009, 383).

Land reform under Corazon Aquino, 1986–92

The early years of the presidency of Corazon Aquino (1986–92), in the immediate aftermath of the overthrow of the Marcos dictatorship, presented a unique opportunity for action on land reform. "There was ample evidence during the government's first year in office that a broad spectrum of opinion favored rapid and decisive action to implement a redistributive program" (Putzel 1992, 213). Sister Crescencia Lucero, executive director of Task Force Detainees of the Philippines, recalled the unfulfilled promise of the heady days of the early Aquino administration (Lucero 2007). From February 1986 until January of 1987, there was a very short democratic space and during this time there was optimism that profound change – including meaningful land reform – would be implemented. However, the 22 January 1987 killing of 13 unarmed farmers protesting for land reform on the Mendiola Bridge near the Malacañang presidential palace by security forces abruptly closed this democratic space. Sister Crescencia recalled how on three separate occasions she met with Corazon Aquino but during the last two meetings, secretary of defense Juan Ponce Enrile and AFP chief of staff General Fidel Ramos were present; two men who were instrumental in keeping Ferdinand Marcos in power were performing the same role for Corazon Aquino. It was during this time that Aquino was presented with – and missed – perhaps the best opportunity to implement genuine land reform capable of reducing rural poverty and unrest (Nadeau 2008).

On 25 May 1986, acting to hasten the reconstruction of a democracy in the islands, Aquino established the Constitutional Commission and entrusted it with the task of writing a new constitution. However, most of the members of this commission were members of the oligarchic ruling class and this revealed Aquino's goal of reestablishing the old order existing prior to martial law (Nadeau 2008). Out of this Constitutional Commission came the 1987 Constitution and Article XIII, Section 4 contained the provision dealing with agrarian reform:

> Section 4. The State shall, by law, undertake an agrarian reform program founded on the right of farmers and regular farmworkers who are landless, to own directly or collectively the lands they till or, in the case of other farmworkers, to receive a just share of the fruits thereof. To this end, the State shall encourage and undertake the just distribution of all agricultural lands, subject to such priorities and reasonable retention limits as the Congress may prescribe, taking into account ecological, developmental, or equity considerations, and subject to the payment of just compensation. In determining retention limits, the State shall respect the right of small landowners. The State shall further provide incentives for voluntary land-sharing.

This section was filled with numerous loopholes[4] allowing the government to forestall the implementation of meaningful land reform but one is readily apparent: the words, "The State shall, by law, undertake an agrarian reform program." The words "by law" meant that Congress – an institution dominated by the landed oligarchy – was to be given "the authority to determine three of the most important characteristics of any reform program: scope of coverage, timing, and phasing, as well as the amount of land that owners would be allowed to retain" (Putzel 1992, 207). To Nadeau (2008, 96) this allowed "for land reform opponents to define the scope for land reform." As Bello et al. (2009, 36) stated:

> Aquino had a rare opportunity to reform the feudal system in the countryside: she enjoyed massive popular support, especially among the middle class; the opposition, particularly the landed bloc, was not yet consolidated; and she possessed 'law-making' powers as head of a 'revolutionary government.' These conditions could have empowered

---

4 See Putzel (1992, 206–10) for a detailed discussion of these loopholes.

her to act quickly to implement a radical reform program. But Aquino copped out and left it to Congress to decide on the program's final fate.

On 10 June 1988, Aquino signed Republic Act 6657 into effect and implemented the Comprehensive Agrarian Reform Program (CARP). The CARP program has been roundly criticized[5] and many have taken to derisively calling it the "compromised agrarian reform program" since it contains so many loopholes allowing landowners to exempt their lands from its application or to delay its application virtually indefinitely. In the words of Franco and Borras (2009, 209), "The CARP process was dominated by antireform policy currents and marked by nepotism, corruption, repression and the non-participation of several rural social movements."

Land reform under neoliberalism: 2001–11

Since the hegemonic acceptance of the precepts of neoliberalism, particularly during the presidency of Gloria Macapagal-Arroyo (2001–10), the government has exhibited minimal enthusiasm for redistributive land reform (Borras 2007; Franco and Borras 2009). As Bello et al. (2009, 65) wrote:

> The state's bias for free enterprise, liberalization, privatization and debt-driven growth shows that it has chosen to focus the nation's resources toward the international market. Instead of dealing with the centuries-old problem of land, the government has placed itself in the service of powerful international financial institutions like the World Bank.

The (in)adequacy of land reform in the Philippines

A lack of effective control over the agricultural lands used by the rural poor remains a serious problem in the twenty-first century and many Filipino farmers still live a serf-like existence on lands owned by members of the oligarchy. The outcomes under CARP are well below those in other countries and the extent of "redistributive land reform outcome is far below the official claims in government statistics" (Borras 2007, 287). The Department of Agrarian Reform (DAR) (the government agency responsible for the implementation of CARP) cites strong landowner resistance and a lack of funds as the primary reasons for the failure of CARP (PESANTEch 2000). Notwithstanding the fact that the regional trial courts lack jurisdiction over agrarian reform cases, landowners will harass CARP beneficiaries by

---

5 See Putzel (1992, 272–6) for a detailed discussion of the problems inherent in CARP.

filing vexations, civil and criminal cases against them and postponing their occupancy of lands awarded to them by making them contest frivolous lawsuits and criminal prosecutions (Franco and Borras 2009). Threats and acts of violence are also frequently used against farmers to prevent them from attempting to acquire land under CARP (PESANTEch 2000). Field personnel from the DAR are often subjected to death threats from landowners if they pursue land reform too vigorously. Carlos Fortich, Governor of Bukidnon, once told the staff of the DAR that "the only way they would be able to enter his property was if he made fertilizer out of them (Bello et al. 2009, 67). A combination of state and nonstate forces has increasingly violated the human rights of peasant land claimants (Franco and Borras 2009). Until there is meaningful land reform, the peonage imposed on the rural poor by an unjust tenancy system will continue to drive the peasantry to rebellion. "Without a real and far-reaching overhaul of the land tenure system, all other rural programs are bound to be merely cosmetic, their efforts ameliorative and their achievements peripheral and temporary" (Constantino and Constantino 1978, 267). As the eminent Filipino historian Teodoro Agoncillo (1990, 460) wrote: "Lincoln said, during one of the most critical periods of American history, that the nation could not remain half slave and half free. The Philippines, no less, cannot forever remain nine-tenths serf and one-tenth landlord."

The Sustainable Use of Natural Resources to Meet Local Needs

The preservation of biodiversity

The Philippines: A biodiversity hotspot

An important opportunity for natural resources to be used sustainably lies in the tremendous biodiversity potential of the archipelago. The Philippines is a veritable biodiversity hotspot containing many rare species endemic only to the islands (Bravante and Holden 2009; Haribon Foundation and Birdlife International 2001; Heaney and Regalado 1998; Peterson et al. 2000). It is reasonable to think of the Philippines as the Galapagos Islands multiplied ten-fold with more than 510 species of land mammals, birds, reptiles and amphibians existing only in the archipelago (Heaney and Regalado 1998). The archipelago, along with Haiti, is among the top two countries in the world in terms of avian endemism (Balmford and Long 1994). Consider Sibuyan Island in the province of Romblon, where approximately 40 percent of all species are endangered (Heaney and Regalado 1998).

This tremendous biodiversity emanates from the complex biogeography of the archipelago (Haribon Foundation and Birdlife International 2001; Heaney and Regalado 1998). Not only is the Philippines separated from the Asian mainland, Taiwan, Borneo and Sulawesi by deep-water channels (which remained filled with water when sea level fell during the last ice age) several deep-water channels also separate the islands of the archipelago from themselves. This has had the effect of limiting the sharing of species among the members of the archipelago. For example, when one compares the mammalian species on the northern tip of the island of Samar with the mammalian species on the southern tip of the island of Luzon (a distance of only 25 kilometers) one will find that 80 percent of the species are different (Heaney and Regalado 1998). As Haribon Foundation and Birdlife International (2001, 23–4) wrote:

> Almost every major international conservation organization now regards the Philippines as one of the highest priority countries in the world for conservation concerns, because of its enormous biological diversity, the extraordinarily high levels of endemism, the high rate of deforestation and habitat destruction.

The Philippines, along with Indonesia and Papua New Guinea, also constitutes one of the three corners of the Coral Triangle (Figure 8.2), the location of the

Figure 8.2. The Coral Triangle

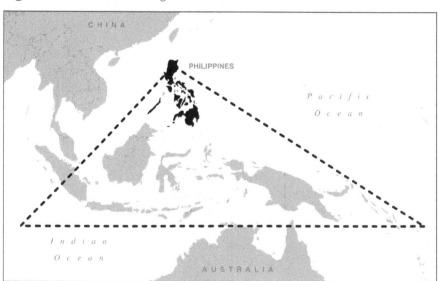

highest coral reef diversity in the world (Regis 2008). The archipelago's coral reefs contain 63 percent of all coral species known worldwide and provide habitat to over 2,000 species of fish (Regis 2008). Should mining continue in the Philippines – particularly in coastal areas such as Rapu-Rapu Island – acid mine drainage and cyanide spills will pose a grave threat to these coral reefs.

The importance of biodiversity

The importance of biodiversity preservation lies in the fact that unique species of life may act as a treasure chest of genetic information capable of producing cures for diseases or providing new sources of food (European Communities 2008). The extinction of even a single endangered species is a tremendous loss to global biodiversity (Haribon Foundation and Birdlife International 2001; Heaney and Regalado 1998). If biodiversity continues to be reduced, irreparable consequences on the world's economy could result.

In Chapter Six, the antimining moratoriums on the island of Samar were discussed. According to Attorney Grizelda Mayo-Anda, another important reason why the Samarnons are opposed to mining is because of the high levels of biodiversity found on the island (Mayo-Anda 2009). Samar is a "biodiversity hotspot," unique for having the country's largest remaining unfragmented tract of lowland tropical rainforest, approximately 360,000 hectares in total (Santos and Lagos 2004). In 2003, the residents of Samar conducted an evaluation of their island's biodiversity resources and concluded that over a 25-year period, it would be worth almost USD 44 billion while, over this same time period, the projected earnings from mineral resource extraction would only be worth USD 21 billion at most (De Alban et al. 2004). In June of 2011, Gina Lopez, managing director of the ABS-CBN Foundation, visited Samar and stated that the residents of that island stand to benefit far more from its rich biodiversity and the promotion of ecotourism than they ever could from mining (Labro 2011).

The preservation of ethnodiversity

Just as the preservation of biodiversity presents an important consideration when the true costs of mining are considered, one must also contemplate the preservation of ethnodiversity. In Chapter Four, the overlap between mining and indigenous peoples was discussed and Figure 4.16 depicted the interaction of mining projects with indigenous peoples by overlaying the location of mining projects with the percentage of each province consisting of indigenous peoples. As Chapter Four indicated, the substantial amount of mining occurring on lands inhabited by indigenous peoples generates concerns about the viability of indigenous cultures.

To the Canadian anthropologist Wade Davis, the preservation of ethnodiversity is as (or possibly, even more) important than the preservation of biodiversity (Davis 2002, 2007, 2009). Just as the earth has a biosphere, which contains all life on our planet, it also has an ethnosphere and this may be defined as "the sum total of thoughts, beliefs, myths, and intuitions brought into being by the human imagination since the dawn of consciousness" (Davis 2002, 57). It may also be conceived of as being "the full complexity and complement of human potential as brought into being by culture and adaptation since the dawn of consciousness" (Davis 2007, 5). Just as biodiversity is threatened by activities which degrade the environment such as mining, ethnodiversity is similarly threatened by such activities when they cause indigenous peoples to be displaced from their ancestral lands. This is because the cultures of indigenous peoples are intimately tied to the particular pieces of land they inhabit (Eder 1987). As Alejo (2000, 20) wrote:

> When we deal with land, we confront something more than abstract space. Land is multidimensional in meaning and valuation. It is a political territory, an economic resource and a cultural and even spiritual base. Marginalization, therefore also means more than location transfer. Those who are displaced due to environmental changes, like deforestation or the construction of big dams, may feel that their whole way of life is being negated. This is particularly true of upland dwelling ethnic minorities who have generally shown a peculiar attachment to land.

As Chapter Four showed, the Igorots of the Cordillera of Luzon are becoming displaced by mining and are moving to Baguio City where they "join the legions of urban poor, living on the edges of a cash economy, trapped in squalor and struggling to survive" (Davis 2007, 199). Once Igorots become displaced by mining, they will lose their attachment to the land from which their culture sprang and they will begin to lose their culture. This loss of ethnodiversity is occurring at a much greater rate – and with more dire consequences – than is the loss of biodiversity. "Tragically, just as the biosphere is being severely eroded, so too is the ethnosphere and at a far greater rate" (Davis 2002, 57). Even "the most apocalyptic assessment of the future of biological diversity scarcely approaches what is known to be the best conceivable scenario for the fate of the world's languages and cultures" (Davis 2007, 5).

Placing a value on this loss of ethnodiversity is, unfortunately, virtually impossible; although, to an extent, protecting ethnodiversity also acts to protect biodiversity. Since many indigenous groups have intimate knowledge about the flora and fauna where they live, these people offer many lessons in resource management in a variety of ecosystems. As the World Commission on Environment and Development (1987, 114–15) wrote:

> These communities are the repositories of vast accumulations of traditional knowledge and experience that link humanity with its ancient origins. Their disappearance is a loss for the larger society which could learn a great deal from their traditional skills in sustainably managing very complex ecological systems.

Consider the Haunoo, forest dwellers from the island of Mindoro, who know the taxonomy of more than 450 species of animals and can differentiate among 1,500 species of plants – 400 species more than those known by Western botanists who work in the same forests (Davies 2007). Since as many as 40,000 species of plants may have medicinal or nutritional properties, the preservation of indigenous cultures also acts to preserve the role of biodiversity as a treasure chest of genetic information (Davies 2007). Beyond this somewhat simplistic example, however, engaging in a valuation of ethnodiversity is virtually impossible other than to say that these cultures are a part of our shared experience as humans and that if they are lost we will move "towards a monochromatic world of monotony from a polychromatic world of diversity" (Davis 2002, 61).

The importance of community-based conservation

A controversy frequently occurring whenever conservation is discussed in the Philippines is the difficulty faced by the managers of protected areas in dealing with those people who rely upon the natural resources in those areas for their subsistence. In countries such as Australia and Canada, with their low population densities (Figure 5.2), protected areas may be managed by a paradigm known as the "fences and fines approach" that "excludes people as residents, prevents consumptive use and minimizes human impacts" (Adams 1998, 301). In the archipelago – with its high population density – this approach cannot be used due to the heavy reliance by the poor and marginalized on natural resources found on public lands. As Utting (2000a, 1) wrote:

> By the late 1980s there was growing recognition among many agencies of the serious limitations of technocratic or authoritarian approaches

to environmental protection that emphasized 'top-down' planning and 'policing' of protected areas and ignored the livelihood concerns, cultural rights, and local knowledge of resource management and ecosystems.

Consequently, the Philippines has become "a country where numerous organizations have been actively involved in promoting more integrated and community-based approaches to natural resources management and protection" (Utting 2000a, 1). These efforts to promote community-based natural resource management have garnered international recognition for the archipelago as a place that is actively pursuing an agenda of participatory conservation and sustainable development (Utting 2000a). Also, in addition to ecological goals, "many community-based resource management projections aim to provide local resource users with tangible human welfare benefits" (Bagadion 2000).

The role of the National Integrated Protected Areas System

The principal vehicle for facilitating community-based conservation in the Philippines is the National Integrated Protected Areas System (NIPAS) established by Republic Act 7586. This statute[6] recognizes that a key ingredient in the effective management of protected areas is the active involvement of local communities living within or adjacent to protected areas and who depend on the resources found there for subsistence (Melgar 1997). The NIPAS provides for a much broader role for these communities to define the appropriate strategies and participate in the conservation and management of the natural resources found in their locality" (Melgar 1997, 127). The people who live off of a natural resource base are ultimately its best protectors, so they should be the ones involved in its conservation.

Evaluating community-based conservation in the Philippines

In the Philippines, "participatory conservation has had a very mixed record" (Utting 2000b, 171). The Mount Makiling Community Based Conservation Program, which was launched in 1992, was arguably a success (Bagadion 2000; Utting 2000b). The 4,244 hectare Mount Makiling Forest Reserve was located in Laguna, approximately 65 kilometers south of Metro Manila. The project

---

6 Section 9 calls for every protected area to have a management plan providing guidelines for the protection of indigenous peoples living within each protected area and for the protection of other communities who have lived continuously within an area for five years prior to its designation as a protected area and who are solely dependent upon that area for subsistence.

consisted of (1) addressing specific local community needs such as lack of access to clean water, flooding and landslides; (2) providing local residents access to health facilities as well as support for the marketing of nontimber forest products such as honey; and (3) clarifying and settling land tenure problems involving forest dwellers. The Puerto Princesa Subterranean River National Park has, however, been less of a success (Dressler 2009). While the park (a United Nations Educational, Scientific and Cultural Organization World Heritage Site) has done an excellent job of protecting the vicinity of the underground river, there is controversy regarding how well the members of the local Tagbanua tribe have fared in its operation. The Tagbanua were denied access to their traditional practices of slash-and-burn agriculture because burnt fields are aesthetically unpleasing to tourists while lowland migrants residing in the park have been allowed to engage in wet rice cultivation. This has generated concerns about the equity with which the park is being operated.

Nevertheless, despite the limitations of participatory approaches to natural resource conservation in the Philippines, these approaches "have often fared much better than their conservative predecessors" (Utting 2000b, 214). As Utting (2000b, 189) wrote:

The long-held assumption of many conservationists that the poor inevitably degrade the environment has been debunked by research in many countries that reveals the positive environmental aspects of 'traditional' resource management systems and so-called indigenous knowledge as well as the myriad forms of collective responses of the rural poor to deal with environmental degradation.

The basic ecclesial community movement: Participatory development

Postmodern public administration

In many places in the developing world, an awareness has emerged that the sustainable use of natural resources to meet local needs can only occur if the poor themselves participate in development. "From the 1970s, calls for 'bottom-up' or 'participatory' approaches to development have grown stronger, with an emphasis on promoting changes that people themselves locally understand to be necessary and desirable" (Adams 1998, 290). Such participatory development is becoming addressed by "postmodern public administration," which is an attempt to "respond to the fundamental needs of the poor and oppressed without necessarily addressing the complexities of political arrangements" (Martin 2003, 83). It can be called "postmodern" because it rejects the trust in experts and the faith in humanly engineered progress, which, as discussed in Chapters Two and Six, are the hallmark of modernity. Rather than placing trust in experts and their scientifically verified theories, it looks for methods to help the poor improve their conditions on their own and attempts are made to improve the conditions of the poor and marginalized without changing power relations in society. Rather than abruptly changing the direction and control of the state, this is an attempt by the poor themselves to bypass the state in seeking remedies for just and sustainable policies. As Vanden (2006, 284) wrote:

> Unlike the radical revolutionary movements of the last decades, these new movements do not advocate the radical restructuring of the state through violent revolution. Rather, their primary focus is to work through the existing political system by pushing it to its limits to achieve necessary change and restructuring.

The basic ecclesial community movement in the Philippines

The Basic Ecclesial Community (BEC) movement organized by the Roman Catholic Church is a good example of participatory development in the Philippines (Holden 2009b; Holden and Nadeau 2010; Nadeau 2002, 2005). The term "basic" refers to both the size and the social location of the BECs. They are small communities, consisting of from 40 to 200 families organized on a parish-by-parish basis, and most BECs consist of the "small people," the poor and the marginalized. Members of all social classes may join BECs but, overwhelmingly, they are a movement of the poor since middle-class and upper-class people have eschewed participation in them. They are called "ecclesial" communities because this emphasizes the place of the BECs within the Roman Catholic Church. They are a way of being a church that is realized, located and experienced at the grassroots. The word "community" emphasizes the communitarian nature of the BECs. These are not societies or associations, but are communities whose members live in close spatial and social proximity to each other and who regularly interact with each other. The role of the BECs as communitarian organizations stands in sharp contrast with the individualistic, selfish, privatized and competitive style marking modern Western culture, particularly under neoliberalism.[7]

---

7 The BECs are intended to be the realization of an idealized and just Christian community. See Acts 2:42 to 2:47 and Acts 4:32 to 4:35.

Sustainable livelihood programs for the poor

The BECs provide a number of sustainable livelihood programs for their members such as handicraft production, food processing, garment making, soap making, cooperative stores, communal farming and livestock rearing programs (Holden 2009b; Holden and Nadeau 2010; Nadeau 2002, 2005). There is a need to address the economic conditions of the poor and the social problems faced by them and this can only be done by responding to the socioeconomic conditions they face. An interesting dimension of BEC livelihood programs is their frequent use of the participatory action research approach wherein the BEC members themselves conduct research on their conditions in life and on what types of livelihood programs they would like to implement. This is an example of community empowerment and its essence is to give the poor a voice and to make them feel that they own the project, as opposed to being mere employees of someone else.

Herbal medicine programs

Given the high cost of pharmaceutical drugs, the low quality of public health care and the biodiversity prevalent in the Philippines, an important component of the BEC movement is the provision of herbal medicine (Holden 2009b; Holden and Nadeau 2010). In the Philippines, even the most rudimentary pharmaceutical drugs are almost always beyond the means of the poor and government hospitals have inadequately stocked pharmacies that do not even carry basic medicines. The biodiversity of the archipelago provides a substantial number of herbal medicines that can act in lieu of prohibitively expensive pharmaceutical drugs. Many BEC members use herbal medicine and members will often gather together to make them. Some BECs have arranged for doctors to provide seminars on how to develop herbal medicine; this serves the two-fold purpose of providing people with something they can sell to others in their *barangay* as well as an affordable form of medicine.

Zones of peace

One method, by which the BECs attempt to improve the conditions of the poor, is by acting on their own (at the level of their own community) to end conflict between the AFP and the NPA by creating zones of peace (Holden 2009b). In a zone of peace, BEC members make it clear that they are autonomous from, and will not be influenced by, either the AFP or the NPA. The BEC members then refuse to give any assistance to the AFP or the NPA; this prevents the former from winning the hearts and minds of the people – a classic counterinsurgency tactic – and it prevents the latter from engaging in mass-based building, so as to pursue the protracted people's war from the countryside. What is interesting about the zones of peace is the fact that both the AFP and the NPA decry their existence claiming that they obey the zones of peace while the other side does not and, consequently, the other side gains an advantage from their existence while they remain hampered by their existence. The fact that both the AFP and the NPA are so opposed to the zones of peace could well be a sign that they are disrupting the activities of both sides, ultimately making it more difficult for both sides to engage in hostilities.

Grassroots environmentalism

The emphasis of the BECs upon the environment is one of their most important activities (Holden 2009b). This consists of activism against environmentally harmful activities, as well as the implementation of programs designed to protect the environment. The environmental activism of the church is clearly apparent in its staunch opposition to mining (Holden and Jacobson 2007a, 2011) and is also involved in activism regarding logging and hydroelectric dams. In the late 1980s, the BECs in San Fernando, Bukidnon were heavily involved in a campaign against logging. This campaign involved roadblocks to stop logging trucks and the picketing of the DENR office in Malaybalay, Bukidnon. Ultimately, this campaign led to the granting of a total log ban in the province of Bukidnon. In the 1990s, the BECs were instrumental in a campaign against a hydroelectric dam proposed on the Salug River in Zamboanga del Sur. The dam was proposed by the National Irrigation Administration and was to receive financial support from the Japanese International Cooperation Agency. The BECs received support from the Justice and Peace Desk of the Diocese of Pagadian and the Justice and Peace Office of the Japanese Bishops Conference, which ultimately succeeded in convincing the Japanese International Cooperation Agency to withdraw its support for the dam.

The BECs also engage in numerous programs designed to help protect the environment such as organic farming, solid waste management and tree planting (Holden 2009b). Organic farming is one of the most important programs designed for this purpose. Conventional chemical-based farming leads to: indebtedness on the behalf of farmers, a decline in soil fertility, lower quality food and it exposes farmers to numerous chemicals such as arsenic – which are carcinogenic (IBON 2006a). Organic farming is taken so seriously by the church that it established the Inter-Diocesan Sustainable Agriculture Network at Xavier University- Ateneo de Cagayan in Cagayan de Oro where training and seminars are provided on organic farming.

Bottom-up instead of top-down development

The current neoliberal development paradigm ascribed to by the Philippine government could be referred to as a top-down development strategy (Holden and Nadeau 2010; Nadeau 2005). According to this strategy, external agents such as multinational corporations are to develop the archipelago by extracting products to be sold on global markets. This top-down approach "considers progress to be determined largely by global economic forces" (Nadeau 2005, 325). "It implies that economic growth will someday trickle down to benefit the majority of local people by generating the surplus needed to solve their problems" (Nadeau 2005, 325). In contrast to the neoliberal development paradigm, the BECs attempt to provide a bottom-up approach involving the poor in their own development. The BECs try to provide what Broad (1988, 233) calls genuine development as they strive to provide a development process that is not so dependent on the world economy and wherein "people participate in making decisions and planning projects that affect their lives – where inhabitants of an area decided what kind of projects they want and what kinds they can afford." The BECs are "part of a global trend toward the development of grassroots alternatives to the hegemonic discourse, symbols and economic structures of global capitalism" (Nadeau 2005, 337). The BECs are attempting to achieve sustainable development by improving the conditions of the poor without harming the environment upon which they depend for their livelihoods and in a manner that fosters social equity. This allows both intergenerational equity (the necessary condition for sustainability in Table 7.1) and intragenerational equity (the necessary condition for development in Table 7.1) to be met. The BEC movement is an attempt to provide development for the poor at a local level as opposed to having development come to the poor by abruptly including them into the global economy.

There are many who have criticized the BEC movement for not doing enough to help the poor and for being largely stalled at a purely liturgical level (Holden 2009b; Holden and Nadeau 2010). However, these criticisms are largely focused on the rapidity of the program's development not on its direction. The program is largely still a work in progress, but it shows that there are alternatives to corporate driven neoliberal globalization. To echo the motto of the World Social Forum, "another world is possible!"

CONCLUSION

At the outset, this book posed the question, "What are the difficulties inherent in attempting to pursue a mining-based development paradigm in a country beset by natural hazards?" The Philippines is a country plagued by widespread poverty, but also richly endowed with mineral resources. To stimulate the economic development of the country, the government has rigorously promoted large-scale mining by corporations. Mining is, however, an activity with a substantial potential for environmental degradation and the Philippines is a country subjected to numerous natural hazards such as typhoons, earthquakes, tsunamis, volcanoes and El Niño–induced drought. The Philippines is also inhabited by poor people engaged in subsistence activities who are highly vulnerable to any form of environmental degradation. These natural hazards interfere with the environmental effects of mining and worsen the conditions of the poor engaged in subsistence activities thus creating disasters. The situation that transpired after the typhoon caused the tailings spill at the Rapu-Rapu Polymetallic Project in October 2005 will play itself out over and over again, possibly with catastrophic consequences. This is the classic embodiment of a disaster: a hazard that impacts a vulnerable population. A mining-based development paradigm will not generate development lifting the poor out of poverty and it will deprive them of their basic means of survival and generate disasters. To answer the question posed by the title of this book, this is not an example of "digging to development" and this is an example of "digging to disaster."

Using a mining-based development paradigm as a method of accelerating the economic development of the Philippines is highly problematic given these hazards and the vulnerable people that will be affected. In many ways, this exposes one of the most serious problems with mining which is the fact that many places should simply be declared areas where mining is not a suitable land use. In discussing the suitability of mining as a land use on the public lands of the American West, Stephen D'Esposito (president of the mining advocacy NGO Earthworks) lamented the fact that the suitability of a location for mining is never properly discussed. In the controversy over whether or not a mine should be developed on a particular piece of public land, the discussion will polarize around the mining project proponent who values the land for the minerals that lie beneath it, on one hand, and on the other hand, groups who value keeping the land in a protected state (D'Esposito 2005). Since the actual issue of whether or not mining should be allowed on the land cannot be raised, the discussion disintegrates into one of competing scientific and technical studies regarding the potential impacts of the mine. An EIS will be prepared for the mining project and then environmental NGOs will challenge the validity of this EIS in the courts while the mining project proponent argues that it was properly prepared. Rarely, if ever, does the discussion focus directly on the question everyone is really concerned about: is mining the appropriate use of this piece of land?

In the opinion of Engineer Virgilio Perdigon, there are many areas in the Philippines that – from an engineering point of view – are the worst possible places to locate mining (Perdigon 2009). Mining and land use are analogous to human sexual intercourse: "no human has ever been born without their parents having had sex but this does not mean that sex can be done anywhere and at any time" (Perdigon 2009, interview). In this regard, Perdigon echoes Ali (2003, 197) who wrote that:

> Mining companies and governments have to realize that just as [an ore] deposit under New York City would not mean that mining will go forward the same may be true for other places as well. This is where environmental justice arguments may start to creep in, despite the geological determinism of mining in general.

Could better planning of land-use prevent these natural hazards from impacting mines and creating disasters? Could every single province in the Philippines pass a mining moratorium and ban large-scale mining within its jurisdiction? Will the Alternative Mining Bill pass in both houses of the Philippine Congress and be signed into law thus prohibiting the location of large-scale mines in areas susceptible to natural hazards? The answers to this flurry of questions do not engender optimism. First, the archipelago is a place where "little or no consideration is given to natural hazards in economic planning" (Bankoff 2003a, 81). Secondly, the oligarchic elite that dominates Philippine society has an immense vested interest in seeing activities such as large-scale mining continue unabated. This book has examined the political economy of mining in the Philippines. The power of those who control Philippine society is so immense and the rewards they stand to receive from mining so substantial that those who control the reigns of power have no incentive to depart from the aggressive promotion of large-scale corporate mining. As Severino

(2000, 87) pointed out, environmental degradation "in the Philippines cannot be fully understood without considering the distribution of power in society... technical and economic explanations are inadequate, as are the solutions and policies that flow from them, [because] the legal and extralegal means still exist for the wealthy and influential to control access to the country's natural resources." Severino continues by stating, "The state, traditionally the vehicle for narrow segments of society to control sources of wealth, still has not evolved into an autonomous and honest broker among competing interests" (2000, 87). This is not a matter of a lack of will to bring in a more thorough system of planning; rather, it is a demonstration of political will by those with vested interests in keeping the system operating as it currently does (Bravante and Holden 2009). As Broad (1995, 331) wrote:

[The] problem in...the Philippines is not a lack of political will but a political will that represents elite...interests. Policy failure on environmental grounds needs to be grasped for what it is-not an oversight, nor as a faulty judgment. The direction of public policy... is too often shaped, both directly and indirectly, by those with a vested interest in the continued mismanagement of natural resources. In other words, one cannot accurately label these as general policy failures or as mismanaged resources. Rather, they are political successes in managing natural resources for the benefit of the controllers.

The capacity of the state to do anything to improve the welfare of the vast majority of the populace has been severely constrained by the control the oligarchy exerts over the state and this control is particularly powerful when it comes to the exploitation of natural resources (Eder 1999). Resource management policy in the archipelago has always been framed in such a way as to benefit the elite, what Broad and Cavanagh (1993, 51–4) refer to as a "plunder economy." With the elite dominating the management of natural resources so as to benefit themselves, concern about managing these resources in such a way as to reduce the vulnerability of the population is virtually nonexistent. As Bankoff (2003a, 140) wrote, "greater vulnerability is the consequence of human activities that have been undertaken on behalf of national or regional elites who have materially benefited by such actions."

Ultimately, the source of the situation whereby an activity as dangerous as hardrock mining has been located in a country susceptible to natural hazards and populated by vulnerable people is capitalism; disasters are essentially spatial outcomes of the process of contemporary capitalism (Wisner et al. 2004). Specifically, this situation has emerged as a result of neoliberalism, the aggressive new strain of capitalism prevailing in recent years; what

Renique (2006, 37) – writing in a Latin American context – referred to as *capitalismo salvaje* (savage capitalism). Neoliberalism is a paradigm emphasizing the opening up of economies to extractive activities such as mining, which can then extract resources and sell them on foreign markets. Neoliberalism, arguably more than anything else, is what has led to the insertion of such a dangerous activity into a place frequented by natural hazards and populated by vulnerable people.

A mining-based development paradigm in the Philippines will lead to events where typhoons and earthquakes collapse tailings dams or where mining-related water table drawdown aggravates an El Niño–induced drought. Such "disasters triggered by natural hazards are a consequence of development failure as much as failed development is the product of disaster" (Schipper and Pelling 2006, 22). The aggressive promotion of large-scale mining is not something that enables the residents of the archipelago to become more resilient to hazards while also ensuring that development efforts do not increase vulnerability to hazards – the two criteria identified by O'Brien at al. (2006) as the essential components of a well-planned development strategy.

The opposition to large-scale mining from the forces of Philippine civil society reveals a number of dimensions: a deep-seated frustration with an environmental impact assessment process failing to avail meaningful opportunities for public participation, a lack of confidence in technology and a lack of trust in experts. This opposition is also a demonstration of the grassroots rejection of neoliberalism that is spreading around the world. From Argentina's occupied factory workers movement to the Zapatistas of Mexico, neoliberalism has been extensively criticized for prioritizing growth over redistribution and efficiency over equity. As Gray and Mosely (2005, 18) wrote, "The effects of neoliberal reforms in many countries have put poorer people at risk." In the words of Sister Crescencia Lucero, "The poor and marginalized bear the brunt of the costs of neoliberalism while a small number receive its benefits; neoliberalism has brought fear and insecurity to the poor" (Lucero 2007, interview). The residents of Barangay Macambol (Figure 6.1) exemplify this fear and insecurity brought by neoliberalism. To these people, a liberalized mining sector that attracts large-scale mining from foreign corporations is not a source of prosperity. Rather, large-scale mining, through its potential environmental effects, poses a very real danger to their livelihoods. These people are left wondering about what will happen to them should there be a mining related environmental disruption. In many ways, these people suffer from a fear of insecurity generated by the uncertainty coming from such a potentially large change to their way of life.

Given mining's environmental effects and the hazards present in the Philippines, a development paradigm based upon large-scale mining is an example of digging to disaster, not an example of digging to development. With this thought in mind, it is time for the last word. This book has made reference to a number of sources (both written and interview) but the last word will go to Boni Dano, an environmental activist with the Surigao del Norte–based NGO Katawhang Simbahan Alang sa Malamboong Kabuhatan (People of the Church for Developing God's Creation) who stated: "With mining there is nothing but broken promises and broken dreams for the next generation" (Dano 2004, interview).

BIBLIOGRAPHY

Adams, W. M. 1998. "Conservation and Development." In *Conservation Science and Action*, ed. W. J. Sutherland, 286–315. Oxford: Blackwell.

Agoncillo, T. A. 1990. *History of the Filipino People*, 8th ed. Quezon City: Garotech Publishing.

Aguilar, J. L. 2008. "Data Mining." *Newsbreak*, July/September, 63.

Alejo, A. E. 2000. "Generating Energies in Mount Apo: Cultural Politics in a Contested Environment." Quezon City: Ateneo de Manila University Press.

———. 2005. Professor of Development Studies, Ateneo de Davao University. Personal interview, Davao City, Philippines, 13 May.

Al Haq, V. 2005. Chair, Moro Islamic Liberation Front Coordinating Committee on the Cessation of Hostilities. Personal interview, Cotabato City, Philippines, 8 June.

Ali, S. H. 2003. *Mining, the Environment and Indigenous Development Conflicts*. Tuscon: University of Arizona Press.

Alon, O. 2005. Civic Affairs Officer, Armed Forces of the Philippines, Sixth Infantry Division. Personal interview, Cotabato City, Philippines, 8 June.

Alston, P. 2007. *Promotion and Protection of all Human Rights, Civil, Political, Economic, Social, and Cultural Rights, Including the Right to Development: Report of the Special Rapporteur on Extrajudicial, Summary, or Arbitrary Executions*. New York: United Nations.

———. 2009. *Promotion and Protection of all Human Rights, Civil, Political, Economic, Social, and Cultural Rights, Including the Right to Development: Report of the Special Rapporteur on Extrajudicial, Summary, or Arbitrary Executions, Follow-Up to Country Recommendations – Philippines*. New York: United Nations.

Alyansa Tigil Mina. 2010. "Green Groups Call for Mining Moratorium to Mitigate El Niño Impacts." http://www.alyansatigilmina.net/content/story/february2010/green-groups-call-mining-moratorium-mitigate-el-ni%C3%B1o-impacts (accessed 17 June 2010).

Amnesty International. 2006. *Philippines: Political Killings, Human Rights, and the Peace Process*. London: Amnesty International.

Anderson, K. 1998. "Social Risk." *Mining Journal*, 6 March, 16.

Ang Bayan. 2003. "Revolutionary Taxation: A Legitimate Act of Governance." *Ang Bayan* 33, no. 1: 3–4.

———. 2006. "On Advancing Peoples' War in Ilocos-Cordillera." *Ang Bayan* 37, no. 20: 5–6.

———. 2011. "Stop Destructive Large-Scale Mining." *Ang Bayan* 42, no. 1: 1–2.

Arguillas, C. O. 2011a. "Tubay Mayor Says Floods Aggravated by Overflow of Mining Firm's Settling Pond." MindaNews. http://mindanews.com/main/2011/02/10/tubay-mayor-says-floods-aggravated-by-overflow-of-mining-firm%E2%80%99s-settling-pond/ (accessed 27 July 2011).

_____. 2011b. "Q and A with Jorge Madlos: 'We can stop them only to a certain degree.'" MindaNews. http://mindanews.com/main/2011/01/09/q-and-a-with-jorge-madlos-"we-can-stop-them-only-to-a-certain-degree"/ (accessed 27 July 2011).

Arnstein, S. 1969. "A Ladder of Citizen Participation." *American Institute of Planners Journal* 35, no. 4: 216–24.

Australia Conservation Federation. 1994. *Mining and Ecologically Sustainable Development: A Discussion Paper.* Fitzroy: Australian Conservation Foundation.

Auty, R. 1994. "Industrial Policy Reform in Six Large Newly Industrializing Countries: The Resource Curse Thesis." *World Development* 22, no. 1: 11–26.

Auty, R. and A. Warhurst. 1993. "Sustainable Development in Mineral Exporting Countries." *Resources Policy* 19, no. 1: 14–29.

Badilla, A. 2005. Program Coordinator, Apostolate Vicariate of Puerto Princesa. Personal interview, Puerto Princesa City, Philippines, 26 April.

Bagadion, B. C. 2000. "Social and Political Determinants of Successful Community-Based Forestry." In *Forest Policy and Politics in the Philippines: The Dynamics of Participatory Conservation*, ed. P. Utting, 117–43. Quezon City: Ateneo de Manila University Press.

Bakker, K. and G. Bridge. 2006. "Material Worlds? Resource Geographies and the 'Matter of Nature." *Progress in Human Geography* 30, no. 1: 1–23.

Balane, W. I. 2010. Bukidnon Bishop: "Save us from election cheating, El Niño." MindaNews. http://www.mindanews.com/index.php?option=com_content&task=view&id=7772&Itemid=50 (accessed 17 June 2010).

Balisacan, A. M. 2003. "Poverty and Inequality." In *The Philippine Economy: Development, Policies, and Challenges*, ed. A. M. Balisacan, 311–41. Oxford: Oxford University Press.

Ballve, T. 2006. "From Resistance to Offensive: NACLA and Latin America." In *Dispatches from Latin America: On the Frontlines against Neoliberalism*, ed. V. Prashad and T. Ballve, 23–31. Cambridge, Massachusetts: South End Press.

Balmford, A. and A. Long. 1994. "Avian Endemism and Forest Loss." *Nature* 372, no. 6507: 623–4.

Bankoff, G. 1999. "A History of Poverty: The Politics of Natural Disasters in the Philippines." *Pacific Review* 12, no. 3: 381–420.

_____. 2001. "Rendering the World Unsafe: 'Vulnerability' as Western Discourse." *Disasters* 25, no. 1: 19–35.

_____. 2003a. *Cultures of Disaster: Society and Natural Hazard in the Philippines.* London: Routledge.

_____. 2003b. "Constructing Vulnerability: The Historical Natural and Social Generation of Flooding in Metropolitan Manila." *Disasters* 27, no. 3: 224–38.

Bankoff, G. and D. Hilhorst. 2009. "The Politics of Risk in the Philippines: Comparing State and NGO Perceptions of Disaster Management." *Disasters* 33, no. 4: 686–704.

Banua, D. 2005. Director, Natripal. Personal interview, Puerto Princesa City, Philippines, 25 April.

Banzon, J. O. 2004. "Seeing God's Providence through Entrepreneurship." In *BECs in the Philippines: Dream or Reality: A Multi-Disciplinary Reflection*, ed. J. Delgado, M. Gabriel, E. Padilla and A. Picardal, 37–42. Taytay, Rizal: Bukal Ng Tipan.

Bautista, G. M. 2008. "A Safe Hedge." *Newsbreak*, July/September, 34–5.

Bebbington, A., I. Hinojosa, D. H. Bebbington, M. L. Burneo and X. Warnaars. 2008. "Contention and Ambiguity: Mining and the Possibilities of Development." *Development and Change* 39, no. 6: 887–914.

Beck, U. 1992. *Risk Society: Towards a New Modernity.* London: Sage.

Bednarz, R. S. 2006. "Environmental Research and Education in US Geography." *Journal of Geography* 30, no. 2: 237–50.

Bello, W., H. Docena, M. de Guzman and M. L. Malig. 2009. *The Anti-Development State: The Political Economy of Permanent Crisis in the Philippines*. Manila: Anvil Publishing.

Beltran, A. 2007. Public Information Officer, Cordillera Human Rights Alliance. Personal interview, Baguio City, Philippines, 30 May.

Bengwayan, A. 2007. Public Information Officer, Cordillera Peoples Alliance. Personal interview, Baguio City, Philippines, 31 May 2005.

Biersack, A. 2006. "Reimagining Political Ecology: Culture/Power/History/Nature." In *Reimagining Political Ecology*, ed. A. Biersack and J. B. Greenberg, 3–40. London: Duke University Press.

Billanes, E. 2005. Environmental activist, South Cotabato, Cotabato, Sultan Kudarat, Sarangani, General Santos City, and Davao del Sur – Alliance for Genuine Development. Personal interview, Koronadal City, Philippines, 18 May.

Blanco, A. V. 2006. "Local Initiatives and Adaptation to Climate Change." *Disasters* 30, no. 1: 140–7.

Blowers, A. B. 2003. "Inequality and Community and the Challenge to Modernization: Evidence from the Nuclear Oases." In *Just Sustainabilities: Development in an Unequal World*, ed. J. Agyeman, R. D. Bullard and B. Evans, 64–80. London: Earthscan Publications.

Boff, L. 1997. *Cry of the Earth, Cry of the Poor*. Maryknoll, New York: Orbis Books.

Bolanio, S. O. 2005. Sister, Oblates of Notre Dame. Personal interview, Davao City, Philippines, 24 May.

Bordo, M. D., R. D. Dittmar and W. T. Gavin. 2007. "Gold, Fiat Money, and Price Stability." *The B.E. Journal of Macroeconomics* 7, no. 1: 1–29.

Borras, S. M. 2007. *Pro-Poor Land Reform: A Critique*. Ottawa: University of Ottawa Press.

Brandes, C., U. Polom and J. Winsemann. 2011. "Reactivation of basement faults: Interplay of ice-sheet advance, glacial lake formation and sediment loading." *Basin Research* 23, no. 1: 53–64.

Bravante, M. A. and W. N. Holden. 2009. "Going Through the Motions: The Environmental Impact Assessment of Nonferrous Metals Mining Projects in the Philippines." *Pacific Review* 22, no. 4: 523–47.

Bridge, G. 2004. "Mapping the Bonanza: Geographies of Mining Investment in an Era of Neoliberal Reform." *Professional Geographer* 56, no. 3: 406–20.

———. 2007. "Acts of Enclosure: Claim Staking and Land Conversion in Guyana's Gold Fields." In *Neoliberal Environments: False Promises and Unnatural Consequences*, ed. Heynen, N., J. Mccarthy, S. Prudham and P. Robbins, 74–86. London: Routledge.

British Broadcasting Corporation. 2009. "In pictures: Philippines floods." http://news.bbc.co.uk/2/hi/in_pictures/8276374.stm (accessed 18 June 2010).

———. 2011. "In graphics: Fukushima nuclear alert." http://www.bbc.co.uk/news/world-asia-pacific-12726591 (accessed 27 July 2011).

Broad, R. 1988. *Unequal Alliance: The World Bank, the International Monetary Fund, and the Philippines*. Berkley: University of California Press.

Broad, R. and J. Cavanagh. 1993. *Plundering Paradise: the Struggle for the Environment in the Philippines*. Berkley: University of California Press.

Broad, R. 1994. "The Poor and the Environment: Friends or Foes?" *World Development* 22, no. 6: 811–22.

———. 1995. "The Political Economy of Natural Resources: Case Studies of the Indonesian and Philippine Forest Sectors." *Journal of Developing Areas* 29, no. 3: 317–40.

Brown, L. R., G. Gardner and B. Halweil. 1999. *Beyond Malthus: Nineteen Dimensions of the Population Challenge*. New York: Norton.

Brown, W. 2003. "Neoliberalism and the End of Liberal Democracy." *Theory and Event* 7, no. 1: 1–21.

Bruch, C. and M. Filbey. 2002. "Emerging Global Norms of Public Involvement." In *The New "Public": The Globalization of Public Participation*, ed. C. Bruch: 1–15. Washington DC: Environmental Law Institute.

Bryant, R. L. and S. Bailey. 1997. *Third World Political Ecology*. London: Routledge.

Burgos, N. P. 2011. "Iloilo Villagers Protest Mining Operations." *Philippine Daily Inquirer*. http://newsinfo.inquirer.net/breakingnews/regions/view/20110401-328770/Iloilo-villagers-protest-mining-operations (accessed 27 July 2011).

Burke, G. 2006. "Opportunities for Environmental Management in the Mining Sector in Asia." *Journal of Environment and Development* 15, no. 2: 224–35.

Cabalda, M. V. 2004. Chief Science Research Specialist, Mining, Environment and Safety Division, Mines and Geosciences Bureau, Department of Environment and Natural Resources. Personal interview, Quezon City, Philippines, 27 July.

Cabangbang, R. 2007. Major, and spokesman, Armed Forces of the Philippines, Eastern Mindanao Command. Personal interview, Davao City, Philippines, 28 June.

Cabanlit, D. P. 2007. Officer in Charge, Davao Office, Philippine Institute of Volcanology and Seismology. Personal interview, Davao City, Philippines, 17 July.

Caliguid, A. K. F. 2008. NPA Raid Mine, Cell Site, Police Station. *Philippine Daily Inquirer*. http://www.inquirer.net/specialfeatures/thesoutherncampaign/view.php?db=1&article=20081222-179436 (accessed 27 July 2011).

Canadian Dimension. 2011. "A Global Mining Powerhouse." *Canadian Dimension* 45, no. 1: 18.

Caouette, D. 2004. "Preserving Revolutionaries: Armed Struggle in the 21st Century, Exploring the Revolution of the Communist Party of the Philippines." PhD diss., Cornell University.

Carreon, E. B. 2009. "Indigenous Women and Mining." In *Mining and Women in Asia: Experiences of Women Protecting Their Communities and Human Rights against Corporate Mining*, ed. V. Yocogan-Diano, 104–9. Chiang Mai, Thailand: Asia Pacific Forum on Women, Law and Development.

Carroll, J. 1986. *Looking Beyond EDSA*. Quezon City: Human Society.

Catedral, R. Father. 2005. Director, Social Action Center, Diocese of Marbel. Personal interview, Koronadal City, Philippines, 18 May.

Caviedes, C. N. 1985. "Emergency and Institutional Crisis in Peru during El Niño, 1982–1983." *Disasters* 9, no. 1: 70–4.

Central Intelligence Agency. 2011. World Factbook. https://www.cia.gov/library/publications/the-world-factbook (accessed 12 July 2011).

Ciencia, A. N. 2006. *The Philippine Supreme Court and the Mining Act Ruling Reversal*. International Graduate Student Conference Series, East-West Center Working Papers No. 29. Honolulu, Hawaii: East-West Center.

Clements, E. L. 1996. "Bust and Bust in the Mining West." *Journal of the West* 35, no. 4: 40–53.

Coban, A. 2004. "Community-Based Ecological Resistance: The Bergama Movement in Turkey." *Environmental Politics* 13, no. 2: 438–60.

Commission on Human Rights of the Philippines. 2011. *Re: Displacement Complaint of Residents of Didipio, Kasibi, Nueva Vizcaya*. CHR-H-2008-0055 (SPL Report). Resolution CHR (IV) No. A2011-004. Quezon City: Commission on Human Rights of the Philippines.

Communist Party of the Philippines Information Bureau. 2011. "Cagayan Town Mayor and Philippine Army Harass Peasants to Sign Mining Endorsement." http://www.ndfp. net/joom15/index.php/peace-talks-mainmenu-75/grp-ndfp-peace-talks-mainmenu-77/1016-cagayan-town-mayor-and-philippine-army-harass-peasants-to-sign-mining-endorsement (accessed 4 June 2011).

Conde, C. H. 2009. Journalist, *International Herald Tribune*. Personal interview, Quezon City, Philippines, 12 November.

Conejar, D. 2007. Mindanao Coordinator, Task Force Detainees of the Philippines. Interview, Davao City, Philippines, 16 July.

Constantino, R., 1975. *The Philippines: A Past Revisited*. Quezon City: Tala Publishing Service.

Constantino, R. and L. R. Constantino. 1978. *The Philippines: The Continuing Past*. Quezon City: The Foundation for Nationalist Studies.

Corkran, R. E. 1996. "Quality of Life, Mining and Economic Analysis in a Yellowstone Gateway Community." *Society and Natural Resources* 9, no. 2: 143–58.

Coronel, S. S., Y. T. Chua, L. Rimban and B. Cruz. 2007. *The Rulemakers: How the Wealthy and the Well-Born Dominate Congress*. Pasig City, Philippines: Anvil Publishing.

Corpus, V. N. 1989. *Silent Wars*. Quezon City: VNC Enterprises.

Crowson, P. 2007. "The Copper Industry 1945–1975." *Resources Policy* 32, no. 1: 1–18.

Cutter, S. L. 1996. "Vulnerability to Environmental Hazards." *Progress in Human Geography* 20, no. 4: 529–39.

Cutter, S. L., B. J. Boruff and W. L. Shirley. 2003. "Social Vulnerability to Environmental Hazards." *Social Science Quarterly* 84, no. 2: 242–61.

Cutter, S. L., L. A. Johnson, C. Finch and M. Barry. 2007. "The US Hurricane Coasts: Increasingly Vulnerable?" *Environment* 49, no. 7: 8–20.

Danenberg, T., C. Ronquillo, J. de Mesa, E. Villegas and M. Piers. 2007. *Fired From Within: Spirituality in the Social Movement*. Manila: Institute of Spirituality in Asia.

Dano, B. 2004. Environmental activist, Katawhang Simbahan Alangsa Malamboong Kabuhatan (People of the Church for Developing God's Creation). Personal interview, Surigao City, Philippines, 15 July.

David, C. C. 2003. "Agriculture." In *The Philippine Economy: Development, Policies, and Challenges*, ed. A. M. Balisacan and H. Hill, 175–218. Oxford: Oxford University Press.

Davis, S. L. 1997. "Engaging the Community at the Tampakan Copper Project: A Community Case Study in Resource Development with Indigenous People." *Natural Resources Forum* 22, no. 4: 233–43.

Davis, W. 2002. "The Naked Geography of Hope." *Whole Earth* (Spring 2002): 57–61.

———. 2007. *Light at the Edge of the World: A Journey through the Realm of Vanishing Cultures*. Vancouver: Douglas and McIntyre.

———. 2009. The Wayfinders: *Why Ancient Wisdom Matters in the Modern World*. Toronto: House of Anansi Press.

Dawe, D., P. Moya and S. Valencia. 2008. "Institutional, Policy and Farmer Responses to Drought: El Niño Events and Rice in the Philippines." *Disasters* 33, no. 2: 291–307.

De Alban, J. D. T., C. D. C. Bernabe and B. E. de la Paz. 2004. *Analyzing Mining as a Threat to Forests and Sustainable Development*. Quezon City: Haribon Foundation.

De Dios M. Pueblos, J. 2005. Bishop, Diocese of Butuan. Personal interview, Butuan, Philippines, 1 June.

De la Cerna, J. 2005. Mayor, Municipality of Governor Generoso, Province of Davao Oriental. Personal interview, Municipality of Governor Generoso, Philippines, 25 May.

Delgado, K. 2007. Karapatan representative, Southern Mindanao. Interview, Davao City, Philippines, 28 June.

Delica, Z. G. 1993. "Citizenry-Based Disaster Preparedness in the Philippines." *Disasters* 17, no. 3: 239–47.

Department of Environment and Natural Resources Climate Change Office. 2010. *The Philippine Strategy on Climate Change Adaptation.* Quezon City: Department of Environment and Natural Resources Climate Change Office.

De Rivero, O. 2001. *The Myth of Development: the Non-Viable Economies of the 21st Century.* London: Zed Books.

D'Esposito, S. 2005. "Public Perspectives on Mining – 'There Must Be a Way to Do it Right.'" In *Mining in New Mexico: the Environment, Water, Economics, and Sustainable Development,* ed. L. G. Price, D. Bland, V. T. McLemore, and J. M. Barker, 41–5. Socorro, New Mexico: New Mexico Bureau of Geology and Mineral Resources.

Devandera, N. P. 2005. Executive Director, Palawan Council for Sustainable Development. Personal interview, Puerto Princesa City, Philippines, 27 April.

De Vera, B. M. 1999. "Sharing the Mineral Wealth: the Philippines' Fiscal Regime." In *Sustainable Development of Land and Mineral Resources in Asia and the Pacific: National Policy Initiatives and Trends in Mining Taxation,* ed. United Nations Economic and Social Commission for Asia and the Pacific, 259–65. New York: United Nations.

Diamond, L. 1999. *Developing Democracy: Toward Consolidation.* Baltimore: John Hopkins University Press.

Diaz, P. P. 2003. *Understanding Mindanao Conflict.* Davao: Mindanao News and Information Cooperative Center.

Diffenbaugh, N. S. and M. Scherer. 2011. "Observational and Model Evidence of Global Emergence of Permanent, Unprecedented Heat in the 20th and 21st Centuries." *Climatic Change.* http://www.springerlink.com/content/l2371617777412kp/fulltext.pdf (accessed 23 June 2011).

Dobb, E. 2002. "Pennies from Hell: In Montana, the Bill for America's Copper Comes Due." In *Montana Legacy: Essays on History People, and Place,* ed. H. W. Fritz, M. Murphy, and R. R. Swartout Jr., 310–40. Helena: Montana Historical Society Press.

Dobry, R. and L. Alvarez. 1967. "Seismic Failures of Chilean Tailings Dams." *Journal of the Soil Mechanics Foundations Division* 93, no. SM6: 237–60.

Doyle, C. 2009. "Indigenous Peoples and the Millennium Development Goals – 'Sacrificial Lambs' or Equal Beneficiaries?" *International Journal of Human Rights* 13, no. 1: 44–71.

Draper, D. L. and B. Mitchell. 2001. "Environmental Justice Considerations in Canada." *Canadian Geographer* 45, no. 1: 39–98.

Dressler, W. K. 2009. "Resisting Local Inequities: Community-Based Conservation on Palawan Island, the Philippines." In *Agrarian Angst and Rural Resistance in Contemporary Southeast Asia,* ed. D. Caouette and S. Turner, 82–104. London: Routledge.

Economist. 2011. *Pocket World in Figures, 2011 edition.* London: Profile Books.

Eder, J. F. 1987. *On the Road to Tribal Extinction: Depopulation, Deculturation, and Adaptive Well-Being among the Batak of the Philippines.* Berkley: University of California Press.

———. 1999. *A Generation Later: Household Strategies and Economic Change in the Rural Philippines.* Honolulu, Hawaii: University of Hawaii Press.

Ellner, S. 2006. "Venezuela: Defying Globalization's Logic." In *Dispatches from Latin America: On the Frontlines Against Neoliberalism,* ed. V. Prashad and T. Ballve, 92–104. Cambridge, Massachusetts: South End Press.

Ellorin, B. G. 2011. "Big business into environmental destruction valid NPA targets – NDFP." MindaNews. http://mindanews.com/main/2011/01/15/big-business-into-environmental-destruction-valid-npa-targets-%E2%80%93-ndfp/ (accessed 28 July 2011).

Emanuel, K. 2005. "Increasing Destructiveness of Tropical Cyclones Over the Past 30 Years." *Nature* 436, no. 4: 686–8.

————. 2007. *What We Know About Climate Change*. Cambridge, Massachusetts: Massachusetts Institute of Technology Press.

Emel, J. and R. Krueger 2003. "Spoken but not Heard: the Promise of the Precautionary Principle for Natural Resource Development." *Local Environment* 8, no.1: 9–25.

Environmental Science for Social Change. 1999. *Mining Revisited*. Quezon City: Environmental Science for Social Change.

Eraker, H. 2000. "Guerilla assault on Norwegian-owned mine: Company demands military protection." http://www.norwatch.no/index.php?artikkelid=692&back=2#Guerilla%20assault%20on%20%20Norwegian-owned%20mine (accessed 21 August 2008).

Escobar, A. 1996. "Constructing Nature: Elements for a Poststructural Political Ecology." In *Liberation Ecologies: Environment, Development, Social Movements*, ed. R. Peet and M. Watts, 46–68. London: Routledge.

Espuelas, A. B. 2008. "Examining the Capacity of the Philippine Army's Enlisted Corps to Accomplish the Government's Counterinsurgency Strategy: Sharpening the Tool." Masters of Military Art and Science thesis, United States Army Command and General Staff College.

Esquillo, L. 2007. Executive Director, Interface Development Interventions. Personal interview, Davao City, Philippines, 17 July.

Estabillo, A. V. 2006. "South Cotabato Braces for El Niño." MindaNews. http://mindanews.com/index.php?option=com_content&task=view&id=1466&Itemid=50 (accessed 19 June 2010).

————. 2007a. "Erratic Weather Patter Worries South Cotabato Agricultural Executives." MindaNews. http://mindanews.com/index.php?option=com_content&task=view&id=2949&Itemid=50 (accessed 19 June 2010).

————. 2007b. "Group Warns of South Cotabato's Vanishing Water Resources." MindaNews. http://www.mindanews.com/index.php?option=com_content&task=view&id=3488&Itemid=106 (accessed 19 June 2010).

————. 2009a. "Sagittarius Bent on Open Pit Mining." MindaNews. http://www.mindanews.com/index.php?option=com_content&task=view&id=6282&Itemid=160 (accessed 19 June 2010).

————. 2009b. "Mining Areas Watched for Landslides." MindaNews. http://www.mindanews.com/index.php?option=com_content&task=view&id=6389&Itemid=160 (accessed 19 June 2010).

European Communities. 2008. "The Economics of Ecosystems and Biodiversity." Wesseling, Germany: European Communities.

Evans, G. J. Goodman and N. Lansbury, eds. 2001. *Moving Mountains: Communities Confront Mining and Globalization*. Sydney: Otford Press.

Fernandez, E. 2006. "MILF Tells Mining Companies to Leave its Territories." http://news.inq7.net/regions/index.php?index=1&story_id=75079 (accessed 21 August 2008).

Finin, G. A. 2008. "'Igorotism,' Rebellion, and Regional Autonomy in the Cordillera." In *Brokering a Revolution: Cadres in a Philippine Insurgency*, ed. R. Rutten, 77–123. Quezon City: Ateneo de Manila University Press.

Flores, A. 2010. "Power, Food Crises Feared." *Manila Standard Today*. http://www.manilastandardtoday.com/insideNation.htm?f=2010/march/1/nation1.isx&d=2010/march/1 (accessed 18 June 2010).

Fonbuena, C. 2008. "On Shaky Ground." *Newsbreak*, July/September, 28–9.

Forestry Management Bureau, Department of the Environment and Natural Resources. 2010. "Forest Cover by Region." http://forestry.denr.gov.ph/landusereg.htm (accessed: 16 June 2010).

Forsyth, T. 2008. "Political Ecology and the Epistemology of Social Justice." *Geoforum* 39, no. 2: 756–64.

Foundation for Environmental Security and Sustainability. 2007. *A Double Edged Sword? Implications of Mining for Environmental Security in the Philippines*. Falls Church, Virginia: Foundation for Environmental Security and Sustainability.

Francaviglia, R. V. 2004. "Hardrock Mining's Effects on the Visual Environment of the West." *Journal of the West* 43, no. 1: 39–51.

Franco, J. C. and P. N. Abinales. 2007. "Again, They're Killing Peasants in the Philippines: Lawlessness, Murder, and Impunity." *Critical Asian Studies* 39, no. 2: 315–28.

Franco, J. C. and S. M. Borras. 2009. "Paradigm Shift: The 'September Thesis' and Rebirth of the 'Open' Peasant Mass Movement in the Era of Neoliberal Globalization in the Philippines." In *Agrarian Angst and Rural Resistance in Contemporary Southeast Asia*, ed. D. Caouette and S. Turner, 206–26. London: Routledge.

Fraser Institute. 2008. *Fraser Institute Annual Survey of Mining Companies 2007/2008*. Vancouver: Fraser Institute.

———. 2011. *Fraser Institute Annual Survey of Mining Companies 2010/2011*. Vancouver: Fraser Institute.

Freedom House. 2010. "Map of Freedom in the World." http://www.freedomhouse.org/template.cfm?page=363&year=2010 (accessed 18 June 2010).

Friedman, M. 1962. *Capitalism and Freedom*. Chicago: University of Chicago Press.

Friedman, M. and R. Friedman. 1979. *Free to Choose: A Personal Statement*. New York: Avon Books.

Fujikura, R. and M. Nakayama. 2001. "Factors Leading to an Erroneous Impact Assessment: A Post Project Review of the Calaca Power Plant, Unit Two." *Environmental Impact Assessment Review* 21, no. 2: 181–200.

Fukuyama, F. 1989. "The End of History." *The National Interest* Summer 1989: 3–18.

Futures Group International. 2004. *Engineering, Health and Environmental Issues Related to Mining on Marinduque: Final Report of the Independent Assessment Team*. Washington DC: The Futures Group International.

Galgana G. A. and M. W. Hamburger. 2010. "Geodetic Observations of Active Intraplate Crustal Deformation in the Wabash Valley Seismic Zone and the Southern Illinois Basin." *Seismological Research Letters* 81, no. 5: 699–714.

Galula, D. 1964. *Counterinsurgency Warfare: Theory and Practice*. London: Praeger Security International.

Garb, P. and G. Komarova. 2001. "Victims of 'Friendly Fire' at Russia's Nuclear Weapons Sites." In *Violent Environments*, ed. N. L. Peluso and M. Watts, 39–65. Ithaca, New York: Cornell University Press.

Garcia, B. 2006. "Zamboanga proposed mining site prone to flooding and landslide." MindaNews. http://www.mindanews.com/2006/03/16nws-mining.htm (accessed 16 March 2006).

Garganera, J. V. 2009. National Coordinator, Alyansa Tigil Mina. Personal interview, Quezon City, Philippines, 3 November.

Gatmaytan, D. 1993. "'Its Too Early For Conservation.' Tokenism in the Environmental Impact Assessment System of the Philippines." Legal Rights and Natural Resources Center Issue Paper #90-05. Quezon City: Legal Rights and Natural Resources Center.

Gatmaytan, A. 2005. Professor of Anthropology, Ateneo de Davao University. Personal interview, Davao City, Philippines, 27.

Gauthier, H. L. and E. J. Taaffe. 2002. "Three 20th Century 'Revolutions' in American Geography." *Urban Geography* 23, no. 6: 503–27.

Gedicks, A. 2001. *Resource Rebels: Native Challenges to Mining and Oil Corporations.* Cambridge, Massachusetts: South End Press.

George, C. 1999. "Testing For Sustainable Development Through Environmental Assessment." *Environmental Impact Assessment Review* 19, no. 2: 175–200.

Geremia, P. Father. 2005. Tribal Filipino Program Coordinator, Diocese of Kidapawan. Personal interview, Kidapawan, Philippines, 18 May.

Giddens, A. 1990. *The Consequences of Modernity.* Stanford, California: Stanford University Press.

GMA News. 2009. "Military blamed for recent killing of activist in Davao Oriental." http://www.gmanews.tv/story/159374/Military-blamed-for-recent-killing-of-activist-in-Davao-Oriental# (accessed 27 May 2010).

——————. 2010. "Anti-mining village chief shot dead in Mindoro." http://www.gmanews.tv/story/183864/anti-mining-village-chief-shot-dead-in-mindoro (accessed 27 May 2010).

Goodland, R. and C. Wicks. 2008. *Philippines: Mining or Food?* London: The Working Group on Mining in the Philippines.

Gomez, J. 2007. "Communist Rebels Own up to Attack on Australian Mine Firm." *Philippine Daily Inquirer.* http://newsinfo.inquirer.net/breakingnews/nation/view_article.php?article_id=93477 (accessed 10 October 2007).

Gonzalez, E. B. 1994. "Tropical Cyclones and Storm Surges." In *Natural Disaster Mitigation in the Philippines: Proceedings, National Conference on Natural Disaster Mitigation 19–21 October 1994, Quezon City, Philippines,* ed. R. S. Punongbayan, 11–8. Quezon City: Philippine Institute of Volcanology and Seismology.

Gorre, I. 2004. Staff Lawyer, Legal Rights and Natural Resources Center. Personal interview, Quezon City, Philippines, 23 July.

Gray, L. C. and W. G. Moseley. 2005. "A Geographical Perspective on Poverty-Environmental Interactions." *Geographical Journal* 171, no. 1: 9–23.

Grollimund, B. and M. D. Zoback. 2001. "Did Deglaciation trigger intraplate seismicity in the New Madrid Seismic zone?" *Geology* 29, no. 2: 175–8.

Guenther, B. 2008. "The Asian Drivers and the Resource Curse in Sub-Saharan Africa: The Potential Impacts of Rising Commodity Prices for Conflict and Governance in the DRC." *European Journal of Development Research* 20, no. 2: 347–63.

Guisadio, C. 2007. Social Action Coordinator, Bula Parish. Personal interview, General Santos City, Philippines, 27 April.

Guerrero, A. [1970] 2006. *Philippine Society and Revolution.* Quezon City: Aklat ng Bayan.

Haas, J. E. 1978. "The Philippine Earthquake and Tsunami Disaster: A Reexamination of Behavioral Propositions." *Disasters* 2, no. 1: 3–11.

Habermas, J. 1981. "Modernity versus Postmodernity." *New German Critique* 22, no.1: 3–14.

Hammes, D. and D. Wills. 2005. "Black Gold: the End of Bretton Woods and the Oil-Price Shocks of the 1970s." *Independent Review* 9, no. 4: 501–11.

Haribon Foundation and Birdlife International. 2001. *Key Conservation Sites in the Philippines.* Makati City: Bookmark.

Harvey, D. 1990. *The Condition of Postmodernism*. Cambridge: Blackwell.

_____. 2003. *The New Imperialism*. Oxford: Oxford University Press.

_____. 2005. *A Brief History of Neoliberalism*. Oxford: Oxford University Press.

_____. 2006. "Neoliberalism as Creative Destruction." *Geografiska Annaler B* 88, no. 2: 145–58.

Hasting, J. P. and K. Mortela. 2008. "The Strategy-Legitimacy Paradigm: Getting it Right in the Philippines." Thesis, Naval Postgraduate School, Monterey, California.

Hatcher, P. 2010. *Investment-Risk in the Philippines: Multilateral Mining Regimes, National Strategies and Local Tensions*. Graduate School on International Relations Working Papers, Faculty of International Relations, Ritsumeikan University. Kyoto: Ritsumeikan University.

Hawes, G. 1987. *The Philippine State and the Marcos Regime: The Politics of Export*. Ithaca, New York: Cornell University Press.

Hayek, F. A. von. 1944. *The Road to Serfdom*. Chicago: University of Chicago Press.

Heaney, L. R. and J. C. Regalado. 1998. *Vanishing Treasures of the Philippine Rain Forest*. Chicago: The Field Museum.

Helmer, M. and D. Hilhorst. 2006. "Natural Disasters and Climate Change." *Disasters* 30, no. 1: 1–4.

Heynen, N., J. Mccarthy, S. Prudham and P. Robbins. 2007. "Conclusion: Unnatural Consequences." In *Neoliberal Environments: False Promises and Unnatural Consequences*, ed. N. Heynen, J. Mccarthy, S. Prudham and P. Robbins, 297–91. London: Routledge.

Hilson, G. 2000. "Barriers to Implementing Cleaner Technologies and Cleaner Production Practices in the Mining Industry: A case study of the Americas." *Minerals Engineering* 13, no. 7: 699–717.

Hinde, C. 2004. "State of the Industry." *Mining Journal*, 13 February, 1.

_____. 2006. "At Risk." *Mining Journal*, 20 October, 2.

Hines, R. K. 2002. "First to Respond to Their Country's Call: The First Montana Infantry and the Spanish American War and Philippine Insurrection, 1898–1899." *Montana: The Magazine of Western History* 52, no. 3: 44–57.

Holden, W. N. 2005a. "Indigenous Peoples and Nonferrous Metals Mining in the Philippines." *Pacific Review* 18, no. 3: 417–38.

_____. 2005b. "Civil Society Opposition to Nonferrous Metals Mining in the Philippines." *Voluntas* 16, no. 3: 223–49.

_____. 2009a. "Ashes from the Phoenix: State Terrorism and the Party-List Groups in the Philippines." *Contemporary Politics* 15, no. 4: 377–93.

_____. 2009b. "Post Modern Public Administration in the Land of Promise: The Basic Ecclesial Community Movement of Mindanao." *Worldviews: Environment, Culture, Religion* 13, no. 2: 180–218.

_____. 2010. "Civil Society." In *Encyclopedia of Human Geography*, ed. B. Warf, 414–16. Thousand Oaks, California: Sage.

_____. 2011. "A Lack of Faith in Technology? Civil Society Opposition to Large-Scale Mining in the Philippines." *International Journal of Science in Society* 2, no 2: 274–99.

Holden, W. N., and A. A. Ingelson. 2007. "Disconnect Between Philippine Mining Investment Policy and Indigenous Peoples' Rights." *Journal of Energy & Natural Resources Law* 25, no. 4: 375–91.

Holden, W. N. and R. D. Jacobson. 2006. "Mining Amid Decentralization: Local Governments and Mining in the Philippines." *Natural Resources Forum* 30, no. 3: 188–98.

_____. 2007a. "Ecclesial Opposition to Nonferrous Metals Mining in the Philippines: Neoliberalism Encounters Liberation Theology." *Asian Studies Review* 31, no. 2: 133–54.

_____. 2007b. "Mining Amid Armed Conflict: Nonferrous Metals Mining in the Philippines." *Canadian Geographer* 51, no. 4: 475–500.

Holden, W. N. and R. D. Jacobson. 2011. "Ecclesial Opposition to Nonferrous Metals Mining in Guatemala and the Philippines: Neoliberalism Encounters the Church of the Poor." In *Engineering the Earth: The Impacts of Megaengineering Projects*, ed. S. Brunn, 383–411. Dordrecht, Netherlands: Kluwer.

Holden, W. N., R. D. Jacobson and K. Moran. 2007. "Civil Society Opposition to Nonferrous Metals Mining in Montana." *Voluntas* 18, no. 3: 266–92.

Holden, W. N. and K. M. Nadeau. 2010. "Philippine Liberation Theology and Social Development in Anthropological Perspective." Philippine Quarterly of Culture and Society. 38, no. 2: 89–129.

Holden, W. N., K. M. Nadeau and R. D. Jacobson. 2011. "Exemplifying Accumulation by Dispossession: Mining and Indigenous Peoples in the Philippines." *Geografiska Annaler: Series B, Human Geography*, 93, no. 2: 141–61.

Holden, W. N. and A. L. Norman 2009. "Public Response to a Student Study of the Feasibility of Nuclear Power: Distrust of Science in a Postmodern World." *International Journal of Science in Society* 1, no. 1: 79–91.

Human Development Network. 2009. *Philippine Human Development Report 2008/2009: Institutions, Politics and Human Development.* Quezon City: Human Development Network.

Human Rights Now. 2008. *Report on Extrajudicial Killings and Enforced Disappearances in the Philippines: Fact Finding Mission of Human Rights Now to Philippines.* Tokyo: Human Rights Now.

Human Rights Watch. 2007. *The Philippines: Scared Silent: Impunity for Extrajudicial Killings in the Philippines.* New York: Human Rights Watch.

_____.2009. *You Can Die Any Time: Death Squads in Mindanao.* New York: Human Rights Watch.

Hutchcroft, P. D. 2008. "The Arroyo Imbroglio in the Philippines." *Journal of Democracy* 19, no. 1: 141–55.

IBON. 2001a. *Birdtalk: Economic and Political Briefing*, 11 January. Manila: IBON Foundation.

_____. 2001b. *Birdtalk: Economic and Political Briefing*, 11 July. Manila: IBON Foundation.

_____. 2002a. *Birdtalk: Economic and Political Briefing*, 15 January. Manila: IBON Foundation.

_____. 2002b. *Birdtalk: Economic and Political Briefing*, 10 July. Manila: IBON Foundation.

_____. 2002c. *IBON Philippines Profile.* Manila: IBON Foundation.

_____. 2003a. *Birdtalk: Economic and Political Briefing*, 16 January. Manila: IBON Foundation.

_____. 2003b. *Birdtalk: Economic and Political Briefing*, 15 July. Manila: IBON Foundation.

_____. 2004a. *Birdtalk: Economic and Political Briefing*, 12 January. Manila: IBON Foundation.

_____. 2004b. *Birdtalk: Economic and Political Briefing*, 15 July. Manila: IBON Foundation.

_____. 2005. *Birdtalk: Economic and Political Briefing*, 13 January. Manila: IBON Foundation.

_____. 2006a. *The State of the Philippine Environment.* Quezon City: IBON Foundation.

_____. 2006b. *Stop The Killings in the Philippines*, 2nd ed. Quezon City: IBON Foundation.

Ilagan, L. 2009. "Legislative Actions on the Mining Issue in the Philippines." In *Mining and Women in Asia: Experiences of Women Protecting Their Communities and Human Rights against Corporate Mining*, ed. V. Yocogan-Diano, 115–23. Chiang Mai, Thailand: Asia Pacific Forum on Women, Law and Development.

Ingelson, A., A. Urzua and W. N. Holden. 2006. "Mine Operator Liability for the Spill of an Independent Contractor in Peru." *Journal of Energy & Natural Resources Law* 24, no. 1: 53–65.

Ingelson, A., W. N. Holden and M. A. Bravante. 2009. "Philippine Environmental Impact Assessment, Mining, and Genuine Development." *Law, Environment, and Development Journal* 5, no. 1: 1–15.

Inni, A. 2005. Advocacy Staff Member, Alternate Forum for Research in Mindanao. Personal interview, Davao City, Philippines, 10 May.

Inocencio, M. 2005. Chairperson, Haribon Palawan. Personal interview, Puerto Princesa City, Philippines, 29 April.

International Centre for Human Rights and Democratic Development. 2007. *Human Rights Impact Assessment for Foreign Investment Projects: Learning from Community Experiences in the Philippines, Tibet, and the Democratic Republic of the Congo, Argentina and Peru*. Montreal: International Centre for Human Rights and Democratic Development.

International Coordinating Secretariat of the Permanent People's Tribunal and IBON Books. 2007. *Repression and Resistance: The Filipino People vs. Gloria Macapagal-Arroyo, George W. Bush et al.* Quezon City: IBON Books.

Isip, E. 2005. Director, Southeast Mindanao Protected Areas and Wildlife Bureau, Department of Environment and Natural Resources. Personal interview, Davao City, Philippines, 3 June.

Justice, Peace and Integrity of Creation Commission-Association of Major Religious Superiors of the Philippines. 2009. *Press Release: What there is a Need for an Alternative Mining Bill*. Quezon City: Justice Peace and Integrity of Creation Commission-Association of Major Religious Superiors of the Philippines.

Joint Foreign Chambers of the Philippines. 2010. *Arangkada Philippines 2010: A Business Perspective*. Manila: Joint Foreign Chambers of the Philippines.

Kalikasan. 2006. "Kalikasan-PNE Pays Tribute to Slain Colleagues; No Justice in Sight for 18 Environmental Activists Killed under Arroyo Administration." Quezon City: Kalikasan.

———. 2011. "Relatives and Friends of Martyred Environmental Activists Call on President Aquino to Immediately Solve Killings." http://www.kalikasan.org/cms/index.php?q=node/386 (accessed 27 May 2011).

Kalyvas, S. N. 2006. *The Logic of Violence in Civil War*. Cambridge: Cambridge University Press.

Karapatan. 2010. *Karapatan 2010 Report on the Human Rights Situation in the Philippines*. Quezon City: Karapatan.

Karnow, S. 1989. *In Our Image: America's Empire in the Philippines*. New York: Ballantine Books.

Katz. R. W. 2002. "Sir Gilbert Walker and a Connection between El Niño and Statistics." *Statistical Science* 17, no. 1: 97–112.

Keynes, J. M. 1936. *The General Theory of Employment, Interest and Money*. Basingstoke: Palgrave Macmillan.

Kirk, D. 2005. *Philippines in Crisis: US Power Versus Local Revolt*. Manila: Anvil Publishing.

Krahmann, E. 2003. "National, Regional, and Global Governance: One Phenomenon or Many?" *Global Governance* 9, no 3: 323–46.

Kuhn, S. 1999. "Expanding Public Participation is Essential to Environmental Justice and the Democratic Decision Making Process." *Ecology Law Quarterly* 25, no. 4: 647–59.

Kuipers, J. R. 2005. "The Environmental Legacy of Mining in New Mexico." In *Mining in New Mexico: the Environment, Water, Economics, and Sustainable Development*, ed. L. G. Price, D. Bland, V. T. McLemore and J. M. Barker, 46–9. Socorro, New Mexico: New Mexico Bureau of Geology and Mineral Resources.

Labonne, B. 1999. "The Mining Industry and the Community: Joining Forces For Sustainable Social Development." *Natural Resources Forum* 23, no. 4: 315–22.

Labro, V. 2010. "Inquirer Visayas: Mining not a top campaign issue in Samar." *Philippine Daily Inquirer.* http://services.inquirer.net/print/print.php?article_id=20100508-268696 (accessed 9 June 2011.).

Labro, V. 2011. "E. Samar Folk Urged: Promote Tourism, Not Mining." *Philippine Daily Inquirer.* http://newsinfo.inquirer.net/14706/e-samar-folk-urged-promote-tourism-not-mining (accessed 5 August 2011).

Lacorte, G. 2011a. "Slain Anti-Mining Activist Knew Threat to His Life, Says Wife." MindaNews. http://mindanews.com/main/2011/04/17/slain-anti-mining-activist-knew-threat-to-his-life-says-wife/ (accessed 27 May 2011).

———. 2011b. "Death Rumors Worry Antimining Activist." *Philippine Daily Inquirer.* http://newsinfo.inquirer.net/8809/death-rumors-worry-anti-mining-activist (accessed 5 August 2011).

Landingin, R. 2008. "Unearthing Strife: Rising Metal Prices and Friendly Government Policy Spur a Surge in Investments and set off a Spate of Conflicts. Protracted Disputes May Hold Back the Industry." *Newsbreak*, July/September, 5–11.

Landingin, R. and J. Aguilar. 2008. "Dirty Past: Many of the Companies Carrying Out the Biggest and Most Important Mining Projects Today Have a History of Releasing Harmful Wastewaters and Substances into the Environment." *Newsbreak*, July/September, 17–19.

Lansang, L. G. F. 2011. "NGOs, Coalition Building and the Campaign for a Minerals Management Policy in the Philippines." *Philippine Political Science Journal* 32, no. 55: 127–66.

Lat, R. A. 1994. "Insurance as a Tool in Disaster Mitigation." In *Natural Disaster Mitigation in the Philippines: Proceedings, National Conference on Natural Disaster Mitigation 19–21 October 1994, Quezon City, Philippines*, ed. R. S. Punongbayan, 175–80. Quezon City: Philippine Institute of Volcanology and Seismology.

Le Billon, P. 2004. "The Geopolitical Economy of 'Resource Wars.'" *Geopolitics* 9, no 1: 1–28.

———. 2005. *Fuelling War: Natural Resources and Armed Conflict.* Adelphi Paper 373. London: The International Institute for Strategic Studies.

Legal Rights Center. 2007. *Antimining Advocate Slain in Surigao del Sur, Legal Rights Center Press Release.* Cagayan de Oro: Legal Rights Center.

Legaspi, R. 2007. "Organizing in the EPZs: The Southern Tagalog Experience." In *Jobs and Justice: Globalization, Labor Rights and Workers' Resistance.* ed. A. A. Tujan, 101–9. Quezon City: IBON Books.

Leonen, M. M. V. F. 2000. "Weaving Worldviews: Possibilities for Empowerment Through Constitutional Interpretation." In *Lawyering for the Public Interest: First Alternative Law Conference*, ed. M. M. V. F. Leonen, 18–36. Quezon City: Alternative Law Groups.

Lewis, J. 2010. "Nevada's Golden Child: Is the State's Hardrock Mining Industry Losing Its Grip?" *High Country News* 42, no. 7: 12–19.

Lim, F. 2011. "Rebels Disarm Mining Firm's Security Men, Kill One." *Philippine Daily Inquirer.* http://newsinfo.inquirer.net/breakingnews/regions/view/20110512-336108/Rebels-disarm-mining-firms-security-men-kill-one (accessed 15 May 2011).

Lima, G. A. C, and S. B. Suslick. 2006. "Estimating the Volatility of Mining Projects Considering Price and Operating Cost Uncertainties." *Resources Policy* 31, no. 1: 86–98.

Linn, B. M. 1989. *The US Army and Counterinsurgency in the Philippine War, 1899–1902.* Chapell Hill, North Carolina: University of North Carolina Press.

Linn, B. M. 2000. *The Philippine War: 1899–1902.* Lawrence: University of Kansas Press.

Lopez, S. P. 1992. *Isles of Gold: A History of Mining in the Philippines.* Oxford: Oxford University Press.

Lucero, C. 2007. Sister, Executive Director, Task Force Detainees of the Philippines. Personal interview, Quezon City, Philippines, 29 May .

Luna, E. M. 2001. "Disaster mitigation and preparedness: The case of NGOs in the Philippines." *Disasters* 25, no. 3: 216–26.

Magalang, L. 2009. Program Officer for Disaster Risk Reduction, Oxfam Great Britain. Personal interview, Quezon City, Philippines, 6 November.

Maglambayan, V. B. 2001. "What Ails the Philippine Minerals Industry? *Public Policy* 5, no. 2: 75–88.

Malone, M. P. and R. B. Roeder. 1976. *Montana: A History of Two Centuries.* Seattle: University of Washington Press.

Mangao, E. A., R. S. G. Gonzales and O. R. Tesoro. 1994. "Seismicity of Mindanao." In *Natural Disaster Mitigation in the Philippines: Proceedings, National Conference on Natural Disaster Mitigation 19–21 October 1994, Quezon City, Philippines,* ed. R. S. Punongbayan, 39–44. Quezon City: Philippine Institute of Volcanology and Seismology.

Manila Observatory. 2010. Mapping Philippine Vulnerability to Environmental Disasters. http://www.observatory.ph/vm/maps.html (accessed 16 June 2010).

Manuel, E. 2004. Staff Lawyer, Legal Rights and Natural Resources Center. Personal interview, Quezon City, Philippines, 12 July.

Mara, S. and S. N. Vlad. 2009. "Natural and Technological Risk Management by Private Insurance in Romania, Including Mining Related Disasters." In *Abstracts of the International Water Conference,* ed. Cilla Taylor Conferences, 967–74. Pretoria, South Africa: Cilla Taylor Conferences.

Mao, Z. [1937] 2005. *On Guerrilla Warfare.* Mineola, New York: Dover Publications.

Martin E. J. 2003. "Liberation Theology, Sustainable Development, and Postmodern Public Administration." *Latin American Perspectives* 30, no. 4: 69–91.

Mayo-Anda, G. 2000. "Case Studies on Mining and EIA: The Palawan Experience." In *Lawyering for the Public Interest: First Alternative Law Conference,* ed. M. M. V. F. Leonen, 226–37. Quezon City: Alternative Law Groups.

———. 2005. Executive Director, Environmental Legal Assistance Center. Personal interview, Puerto Princesa City, Philippines, 21 April.

———. 2009. Executive Director, Environmental Legal Assistance Center. Personal interview, Puerto Princesa City, Philippines, 1 December.

Maza, C. C. 2002. Member, Frente de Defensor de Tambogrande. Personal interview, Tambogrande, Peru, 22 August.

McCarthy, J. 2007. "Privatizing Conditions of Production: Trade Agreements as Neoliberal Environmental Governance." In *Neoliberal Environments: False Promises and Unnatural Consequences,* ed. N. Heynen, J. Mccarthy, S. Prudham and P. Robbins, 38–50. London: Routledge.

McColl, R. W. 1969. "The Insurgent State: Territorial Bases of Revolution." *Annals of the Association of American Geographers* 59, no. 4: 613–31.

McCoy, A. W. 2009. *Policing America's Empire: The United States, the Philippines, and the Rise of the Surveillance State*. Madison: University of Wisconsin Press.

McIlwain, A. I. B. 2003. "Managing Risk at Rapu-Rapu, the Philippines." In *Mining Risk Management Conference: Effective Risk Management for Mining Project Optimization, 9–12 September 2003, Sydney, New South Wales*, ed. Mining Risk Management Conference, 1–9. Carlton, Victoria: Australasian Institute of Mining and Metallurgy.

McLaughlin, D. H. 1956. "Man's Selective Attack on Ores and Minerals." In *Man's Role In Changing The Face Of The Earth*, ed. W. L. Thomas, 851–61. Chicago: University of Chicago Press.

McMahon, G., J. L. Evia, A. Pasco-Font and J. M. Sanchez. 1999. *An Environmental Study of Artisanal, Small, and Medium Mining in Bolivia, Chile, and Peru*. World Bank Technical Paper No. 429. Washington DC: World Bank.

Melecio, R. 2005. Mindanao Regional Coordinator, Task Force Detainees of the Philippines. Personal interview, Davao City, Philippines, 24 May.

Melgar, M. A. 1997. "Shareholders in the Environment: A Case Study of the NGOs for Integrated Protected Areas." In *Civil Society Making Civil Society*, ed. M. C. Ferrer, 127–48. Quezon City: Third World Studies Center.

Melo Commission. 2007. *Independent Commission to Address Media and Activist Killings Created Under Administrative Order No. 157 (s. 2006)*. Manila: Mello Commission.

Mero, S. 2007. Deputy Secretary General, Cordillera Peoples Alliance. Personal interview, Baguio City, Philippines, 31 May.

MindaNews. 2006. "Geohazard Mapping Confirms Tampakan Mining Area Prone to Landslides." MindaNews. http://www.mindanews.com/2006/05/12nws-mapping.htm (accessed 12 May 2006).

———. 2008. "Earthquake Swarm Hits North Cotabato." MindaNews. http://www.mindanews.com/index.php?option=com_content&task=view&id=4012&Itemid=50 (accessed 24 June 2010).

———. 2009. Koronadal Bishop Wants Probe on Killing of Anti-Mining Leader. MindaNews. http://www.mindanews.com/index.php?option=com_content&task=view&id=6062&Itemid=160 (accessed 12 March 2009).

Mineral Policy Institute. 2000. *Cyanide Crash: Report on the Tolukuma Gold Mine Spill of March 2000 in Papua New Guinea*. Sydney: Mineral Policy Institute.

Mines and Geosciences Bureau. 1999. *Chromite Commodity Profile*. Quezon City: Department of Environment and Natural Resources, Mines and Geosciences Bureau.

———. 2000. *Copper: Commodity Profile*. Quezon City: Department of Environment and Natural Resources, Mines and Geosciences Bureau.

———. 2004. *Complete Staff Work: Marcopper Mining Corporation 1996 Tailings Spillage*. Quezon City: Department of Environment and Natural Resources, Mines and Geosciences Bureau.

———. 2006. *Priority Mineral Development Projects of the Philippines*. Quezon City: Department of Environment and Natural Resources, Mines and Geosciences Bureau.

———. 2007a. *Philippine Mining Development Projects Profile*. Quezon City: Department of Environment and Natural Resources, Mines and Geosciences Bureau.

———. 2007b. *Mineral Exploration Projects in the Philippines*. Quezon City: Department of Environment and Natural Resources, Mines and Geosciences Bureau.

———. 2007c. *Priority Mineral Development Projects of the Philippines*. Quezon City: Department of Environment and Natural Resources, Mines and Geosciences Bureau.

Mining Journal. 1993. "Philippines Mining: Heading for Collapse?" *Mining Journal* 320, no. 8210: 89.

———. 1995. "Philippines Gets the Go-Signal." *Mining Journal* 324, no. 8317: 179.

———. 2005. "New Start for the Philippines?" *Mining Journal*, 21 January, 20–5.

———. 2006. "Another Typhoon Hits Rapu-Rapu." *Mining Journal*, 10 November, 5.

Mining, Minerals, and Sustainable Development. 2002. *Breaking New Ground.* London: Earthscan Publications.

Miranda, M., D. Chambers and C. Coumans. 2005. *Framework for Responsible Mining: A Guide to Evolving Standards.* Bozeman, Montana: Center for Science in Public Participation.

Mitchell, A. H. G. and T. M. Leach. 1991. *Epithermal Gold in the Philippines: Island Arc Metallogenesis, Geothermal Systems, and Geology.* London: Academic Press.

Moodie, S. 2001. *Mine Monitoring Manual: A Resource for Community Members.* Whitehorse, Yukon: Yukon Conservation Society.

Moody, R. 2007. *Rocks and Hard Places:The Globalization of Mining.* London: Zed Books.

Morales, F. M. 2008. "Masara Landslide Alarms Environmental Group." MindaNews. http://www.mindanews.com/index.php?option=com_content&task=view&id=5133&Itemid=247 (accessed 25 June 2010).

Morales, N. J. 2010. "Government raises mining investment forecast." http://www.bworldonline.com/main/content.php?id=8793 (accessed 7 August 2011).

Moran, R. 1998. *Cyanide Uncertainties: Observations on the Chemistry, Toxicity, and Analysis of Cyanide in Mining Related Waters.* Washington DC: Mineral Policy Center.

Muradian, R., J. Martinez-Alier and H. Correa. 2003. "International Capital versus Local Population: The Environmental Conflict of the Tambogrande Mining Project, Peru." *Society and Natural Resources* 16, no. 8: 775–92.

Myers, N. 1988. "Environmental Degradation and Some Economic Consequences in the Philippines." *Environmental Conservation* 15, no. 3: 303–11.

Nadeau, K. M. 1992. "Capitalist Southeast Asian Peasants: Some Thoughts on an Old Debate." *Philippine Sociological Review* 40, no. 1: 57–75.

———. 2002. *Liberation Theology in the Philippines: Faith in a revolution.* Westport, Connecticut: Praeger.

———. 2005. "Christians against Globalization in the Philippines." *Urban Anthropology* 34, no. 4: 317–39.

———. 2008. *The History of the Philippines.* Westport, Connecticut: Greenwood Press.

Naito, K., F. Remy and J. P. Williams. 2001. *Review of Legal and Fiscal Framework for Exploration and Mining.* London: Mining Journal Books.

Nally, F. Father. 2007. Columban priest. Personal interview, London, England, 9 November.

Nasol, R. M. 2009. "Albay to Evacuate 47,000 Around Mayon." *Philippine Daily Inquirer.* http://newsinfo.inquirer.net/breakingnews/regions/view/20091215-242075/Albay-to-evacuate-47000-around-Mayon (accessed 3 June 2010).

National Democratic Front of the Philippines. 2007. *The Lies of GRP Officials on Extrajudicial Killings.* Booklet No. 9, NDFP Human Rights Monitoring Committee. Quezon City: National Democratic Front of the Philippines.

National Research Council. 1999. *Hardrock Mining on Federal Lands.* Washington DC: National Academy Press.

National Statistical Coordination Board. 2005. *Estimation of Local Poverty in the Philippines.* Quezon City: National Statistical Coordination Board.

———. 2011. "Poverty Statistics." http://www.nscb.gov.ph/poverty/default.asp (accessed 25 June 2011).

Neame, A. 2005. East Asia Program Manager, Christian Aid. Personal interview, Davao City, Philippines, 6 May.

Neri, R. L. 2005. *Importance of the Mining Sector to the Philippine Economy*. Manila: National Economic Development Authority.

New Internationalist. 2007. *The World Guide: Global Reference, Country by Country*, 11th ed. Oxford: New Internationalist.

Newson, L. A. 1999. "Disease and Immunity in the Pre-Spanish Philippines." *Social Science and Medicine* 48, no. 12: 1833–50.

Obanil, R. 2009. Communications and Networking Officer, Legal Rights and Natural Resources Center. Personal interview, Quezon City, Philippines, 4 November.

O'Brien, G., P. O'Keefe, J. Rose and B. Wisner. 2006. "Climate Change and Disaster Management." *Disasters* 30, no. 1: 64–80.

Obusan, A. C. 2009. Climate and Energy Campaigner, Greenpeace Southeast Asia. Personal interview, Quezon City, Philippines, 4 November.

Ofreneo, R. E. 2009. "Failure to Launch: Industrialization in Metal-Rich Philippines." *Journal of the Asia Pacific Economy* 14, no. 2: 194–209.

Olchondra, R. T. 2011. "PH Misses 2010 Mining Investment Target." http://globalnation. inquirer.net/news/breakingnews/view/20110206-318849/PH-misses-2010-mining-investment-target (accessed 18 April 2011).

Oliveros, R. 2002. *Philippine History and Government*, 2002 ed. Manila: IBON Books.

Orellana, M. A. 2002. "Participation in Minerals Policy." In *The New "Public": The Globalization of Public Participation*, ed. C. Bruch, 235–50. Washington DC: Environmental Law Institute.

Ortolano, L. and A. Shepherd. 1995. "Environmental Impact Assessment: Challenges and Opportunities." *Impact Assessment* 13, no. 1: 3–30.

Oxfam Australia. 2008. *Mining Ombudsman Case Report: Rapu-Rapu Polymetallic Mine*. Carlton Victoria: Oxfam Australia.

Padilla, G. 2007. International Liaison Officer, Karapatan. Personal interview, Quezon City, Philippines, 3 June.

Parreno, E. 2008. "Second Life: a History of Corporate Benevolence is Helping Atlas Mining Win Community Support in Cebu for the Reopening of What Used to be Asia's Biggest Copper Mine." *Newsbreak*, July/September, 49–52.

Parreno, A. A. 2010. *Report on the Philippine Extrajudicial Killings (2001–August 2010)*. San Francisco: The Asia Foundation.

Patalinghug-Vasquez, M. 2005. Central and Eastern Visayas Coordinator, Task Force Detainees of the Philippines. Personal interview, Cebu City, Philippines, 12 June.

Pattison, W. D. 1964. "Four traditions of Geography." *Journal of Geography* 63, no. 5: 211–16.

Paz Luna, M. 2005. *Avenues for People's Participation in the Philippine EIA System*. Paper presented at the EIA in the Philippines, Roads Taken, and Lessons Learned Forum. 11 February 2005. http://web.kssp.upd.edu.ph/eis/fora_proceedings_2005-02-11.html (accessed 25 June 2010).

PEASANTEch. 2000. "Stalled: The Legal Struggles of Farmers for Agrarian Reform." In *Lawyering for the Public Interest: First Alternative Law Conference*, ed. M. M. V. F. Leonen, 126–64. Quezon City: Alternative Law Groups.

Peck, M. J., H. H. Landsberg and J. E. Tilton. 1992. "Introduction." In *Competitiveness inMetals: the Impact of Public Policy*, ed. M. J. Peck, H. H. Landsberg and J. E. Tilton, 1–20. London: Mining Journal Books.

Peet, R. 2003. *Unholy Trinity: the IMF, World Bank and WTO*. London: Zed Books.

Peet, R. and E. Hartwick. 2009. *Theories of Development: Contentions, Arguments, Alternatives.* New York: Guildford Press.

Pegg, S. 2006. "Mining and Poverty Reduction: Transforming Rhetoric into Reality." *Journal of Cleaner Production* 14, nos. 3–4: 376–87.

Penate, G. 2007. Bishop, Apostolic Vicariate of Izabal. Personal interview, Puerto Barrios, Guatemala, 20 February.

Perdigon, V. 2009. Secretary General, Aquinas University; spokesperson for Save Rapu-Rapu Alliance. Personal interview, Legazpi City, Philippines, 11 November.

Perez, A. 2005. Senior Staff Lawyer, Tanggol Kalikasan. Personal interview, Quezon City, Philippines, 20 April .

Peterson, A. T., L. G. Ball and K. W. Brady. 2000. "Distribution of the Birds of the Philippines: Biogeography and Conservation Priorities." *Bird Conservation International* 10, no. 1: 149–67.

Philander, S. G. H. 1990. *El Niño, La Niña and the Southern Oscillation.* San Diego, California: Academic Press.

————.2004. *Our Affair with El Niño. How we Transformed an Enchanting Peruvian Current into a Global Climate Hazard.* Princeton, New Jersey: Princeton University Press.

Philippine Indigenous Peoples Links. 2007. *Chronology of Tailings Dam Failures in the Philippines (1982–2007).* London: Philippine Indigenous Peoples Links.

Philippine Star. 2010. "Malacañang Expresses Full Support For Mining Industry." http:// www.philstar.com/Article.aspx?articleId=642115&publicationSubCategoryId=67 (accessed 15 May 2011).

Picardal, A. L. 1995. *Basic Ecclesial Communities in the Philippines: An Ecclesiological Perspective.* Doctoral dissertation, Faculty of Theology, Pontifical Gregorian University. Rome: Pontifical Gregorian University.

Pinoy Press. 2008. "Anti-mining activist gunned down in Compostela Valley Province." http://www.pinoypress.net/2008/12/24/anti-mining-activist-gunned-down-in-comval/ (accessed 7 August 2011).

Plantilla, A. E. 2005. Executive Director, Haribon Foundation. Personal interview, Quezon City, Philippines, 21 April.

————. 2006. "Nature for Life," *Manila Times.* http://www.abs-cbnnews.com/storypage. aspx?StoryId=29616 (accessed 11 February 2006).

Plumlee, G. S., R. A. Morton, T. P. Boyle, J. H. Medlin and J. A. Centeno. 2000. *An Overview of Mining-Related Environmental and Human Health Issues, Marinduque Island, Philippines: Observations from a Joint U.S. Geological Survey – Armed Forces Institute of Pathology Reconnaissance Field Evaluation, May 12–19, 2000,* Reston Virginia: United States Geological Survey.

Plummer, C. C. and D. McGeary. 1982. *Physical Geology,* 2nd ed. Dubuque, Iowa: William C. Brown.

Poutanen M., E. R. Ivins. 2010. "Upper mantle dynamics and quaternary climate in cratonic areas (DynaQlim)-Understanding the glacial isostatic adjustment." *Journal of Geodynamics* 50, no. 1: 2–7.

Power, T. M. 1996. *Lost landscapes and Failed Economies: The Search for a Value of Place.* Washington DC: Island Press.

————. 2002. *Digging To Development? A Historical Look at Mining and Economic Development.* Washington DC: Oxfam America.

————. 2005. "The Economic Anomaly of Mining – Great Wealth, High Wages, Declining Communities." In *Mining in New Mexico: the Environment, Water, Economics, and Sustainable Development,* ed. L. G. Price, D. Bland, V. T. McLemore and J. M. Barker, 96–9. Socorro, New Mexico: New Mexico Bureau of Geology and Mineral Resources.

_____. 2008. *Metals Mining and Sustainable Development in Central America: An Assessment of Benefits and Costs*. Washington DC: Oxfam America.

Pratt, G. 2008. "International Accompaniment and Witnessing State Violence in the Philippines." *Antipode* 40, no. 5: 751–79.

Pring, G., J. Otto and K. Naito. 1999. "Trends in International Environmental Law Affecting the Minerals Industry." *Journal of Energy & Natural Resources Law* 17, no. 1: 39–56.

Punongbayan, R. S. 1994. "Natural Hazards in the Philippines." In *Natural Disaster Mitigation in the Philippines: Proceedings, National Conference on Natural Disaster Mitigation 19–21 October 1994, Quezon City, Philippines*, ed. R. S. Punongbayan, 1–10. Quezon City: Philippine Institute of Volcanology and Seismology.

Putzel, J. 1992. *A Captive Land: The Policies of Agrarian Reform in the Philippines*. London: Catholic Institute for International Relations.

Pye-Smith, C. 1997. *The Philippines: In Search of Justice*. Oxford: Oxfam UK and Ireland.

Quimpo, F. 2009. "Current Trends in World Mining." In *Mining and Women in Asia: Experiences of Women Protecting Their Communities and Human Rights Against Corporate Mining*, ed. V. Yocogan-Diano, 72–87. Chiang Mai, Thailand: Asia Pacific Forum on Women, Law and Development.

Quimpo, N. G. 2009. "The Philippines: Predatory Regime, Growing Authoritarian Features." *Pacific Review* 22, no. 3: 335–53.

Radetzki, M., R. Eggert, G. Lagos, M. Lima and J. E. Tilton. 2008. "The Boom in Mineral Markets: How Long Might it Last?" *Resources Policy* 33, no. 2: 125–8.

Rapu-Rapu Fact Finding Commission. 2006. *Findings and Recommendations of the Fact Finding Commission on the Mining Operations in Rapu-Rapu Island*. Manila: Rapu-Rapu Fact Finding Commission.

Regis, E. 2008. *Impacts of Mining in an Island Ecosystem: the Case of Rapu-Rapu Island, Philippines*. Naga City, Philippines: Ateneo de Naga University.

_____. 2009. Director, Institute for Environmental Conservation and Research, Ateneo de Naga University. Personal interview, Naga City, Philippines, 9 November.

Renique, G. 2006. "Strategic Challenges for Latin America's Anti-Neoliberal Insurgency." In *Dispatches from Latin America: On the Frontlines against Neoliberalism*, ed. V. Prashad and T. Ballve, 35–46. Cambridge, Massachusetts: South End Press.

Rico, M., G. Benito and A. Diez-Herrero. 2008a. "Floods from Tailings Dam Failures." *Journal of Hazardous Materials* 154, nos. 1–3: 79–87.

Rico, M., G. Benito, A. R. Salgueiro, A. Diez-Herrero and H. G. Pereira. 2008b. "Reported Tailings Dam Failures: A Review of the European Incidents in the Worldwide Context." *Journal of Hazardous Materials* 154, nos. 1–3: 846–52.

Rimando, R. R. 1994. "The Philippine Fault Zone and Hazards due to Faulting." In *Natural Disaster Mitigation in the Philippines: Proceedings, National Conference on Natural Disaster Mitigation 19–21 October 1994, Quezon City, Philippines*, ed. R. S. Punongbayan, 61–70. Quezon City: Philippine Institute of Volcanology and Seismology.

Ripley, E. A., R. E. Redmann and J. Maxwell. 1978. *Environmental Impact of Mining in Canada*. Kingston, Ontario: Center for Resource Studies.

Ripley, E. A., R. E. Redmann and A. A. Crowder. 1996. *Environmental Effects of Mining*. Delray Beach, Florida: St. Lucie Press.

Rodolfo, K. S. and F. S. Siringan. 2006. "Global Sea-Level Rise is Recognized, but Flooding from Anthropogenic Land Subsidence is Ignored Around Northern Manila Bay, Philippines." *Disasters* 30, no. 1: 118–39.

Rodriguez, R.M. 2010. *Migrants For Export: How the Philippine State Brokers Labor to the World*. Minneapolis: University of Minnesota Press.

Rood, S. 1998. "NGOs and Indigenous Peoples." In *NGOs, Civil Society, and the Philippine State*, ed. G. S. Silliman and L. G. Noble, 138–56. Honolulu: University of Hawaii Press.

Rosenau-Tornow, D., P. Bucholz, A. Riemann and M. Wagner. 2009. "Assessing the Long-Term Supply Risks for Mineral Raw Materials – A Combined Evaluation of Past and Future Trends." *Resources Policy* 34, no. 4: 161–75.

Ross, M. 1999. "The Political Economy of the Resource Curse." *World Politics* 51, no. 2: 297–322.

––––––. 2001. *Extractive Sectors and the Poor*. Washington DC: Oxfam America.

Ross, W. A. and D. Thompson. 2002. "Environmental Impact Assessment." In *Tools For Environmental Management*, ed. D. Thompson, 231–45. Gabriola Island, British Columbia: New Society Publishers.

Roque, C. R. and M. I. Garcia. 1993. "The Ecology of Rebellion: Economic Inequality, Environmental Degradation and Civil Strife in the Philippines." *Solidarity* 139, no. 140: 88–120.

Rovillos, R. D., S. B. Ramo and C. Corpus. 2003. "Philippines: When the 'isles of gold' turn to isles of dissent." In *Extracting Promises: Indigenous Peoples, Extractive Industries, and the World Bank*, ed. M. Colchester, A. L. Tamyo, R. Rovillos and E. Caruso, 200–37. Baguio City: Tebtebba Foundation.

Rutten, R. 2008. "Introduction: Cadres in Action, Cadres in Context." In *Brokering a Revolution: Cadres in a Philippine Insurgency*, ed. R. Rutten, 1–34. Quezon City: Ateneo de Manila Press.

Sachs, J. D. and A. M. Warner. 1999. "The Big Push, Natural Resource Booms and Growth." *Journal of Development Economics* 59, no. 1: 43–76.

––––––. 2001. "Natural Resources and Economic Development: The Curse of Natural Resources." *European Economic Review* 45, nos. 4–6: 827–38.

Saldivar-Sali, A. and H. H. Einstein. 2007. "A Landslide Risk Rating System for Baguio, Philippines." *Engineering Geology* 91, no. 1: 85–99.

Sales, P. M. 2009. "State Terror in the Philippines: The Alston Report, Human Rights and Counterinsurgency under the Arroyo Administration." *Contemporary Politics* 15, no. 3: 321–36.

Salomia, M. A. 2005. Father, Pastoral Director, Diocese of Mati. Personal interview, San Isidro, Philippines, 9 May.

Sampat, P. 2003. "Scrapping Mining Dependence." In *State of the World 2003: AWorldwatch Institute Report on Progress Toward a Sustainable Society*, ed. L. Starke, 110–29. New York: Norton.

Samuelson, P. A. 1948. *Economics: An Introductory Analysis*. New York: McGraw Hill.

San Juan, E. 2006. "Neocolonial State Terrorism and the Crisis of Comprador/ Imperialist Hegemony." In *Kontra-Gahum: Academics against Political Killings*, ed. R. B. Tolentino and S. S. Raymundo, 3–27. Quezon City: IBON Books.

Santos, P. V. M. 2010. "The Communist Front: Protracted People's War and Counter-Insurgency in the Philippines (Overview)." In *Primed and Purposeful: Armed Groups and Human Security Efforts in the Philippines*, ed. D. Rodriguez, 17–42. Geneva: Small Arms Survey.

Santos, R. A. M. and B. O. Lagos. 2004. *The Untold People's History: Samar Philippines*. Los Angeles: Sidelakes Press.

Santos, S. M. and P. V. M. Santos. 2010a. "Moro Islamic Liberation Front and its Bangsamoro Islamic Armed Forces." In *Primed and Purposeful: Armed Groups and Human Security Efforts in the Philippines*, ed. D. Rodriguez, 344–61. Geneva: Small Arms Survey.

_____. 2010b. "Communist Party of the Philippines and its New People's Army. In *Primed and Purposeful: Armed Groups and Human Security Efforts in the Philippines*, ed. D. Rodriguez, 261–76. Geneva: Small Arms Survey.

Saturay, J. 2007. Former member, Alliance Against Mining in Oriental Mindoro. Personal interview, Utrecht, Netherlands, 11 November.

Saturay, R. 2009. Geologist and Program Coordinator, Center for Environmental Concerns. Personal interview, Quezon City, Philippines, 5 November.

Sanz, P. 2007. "The Politics of Consent: the State, Multinational Capital, and the Subanen of Canatuan." In *Negotiating Autonomy: Case Studies on Philippine Indigenous Peoples' Land Rights*, ed. A. B. Gatmaytan, 109–35. Copenhagen: International Working Group on Indigenous Affairs.

Schipper, L. and M. Pelling. 2006. "Disaster Risk, Climate Change and International Development: Scope for, and Challenges to, Integration." *Disasters* 30, no. 1: 19–38.

Seck, S. 1999. "Environmental Harm in Developing Countries Caused by Subsidiaries of Canadian Mining Companies: The Interface of Public and Private International Law." *Canadian Yearbook of International Law* 37: 139–221.

Segubiense, R. S, Father. 2009. Executive Director, Social Action Center, Diocese of Legazpi. Personal interview, Legazpi City, Philippines, 10 November.

Severino, H. G. 2000. "The Role of Local Stakeholders in Forest Protection." In *Forest Policy and Politics in the Philippines: The Dynamics of Participatory Conservation*, ed. P. Utting, 84–116. Quezon City: Ateneo de Manila University Press.

Shomaker, J. W. 2005. "Will there be Water to Support Mining's Future in New Mexico?" In *Mining in New Mexico: the Environment, Water, Economics, and Sustainable Development*, ed. L. G. Price, D. Bland, V. T. McLemore and J. M. Barker, 128–30. Socorro, New Mexico: New Mexico Bureau of Geology and Mineral Resources.

Shrader-Frechette, K. S. 1993. *Burying Uncertainty: Risk and the Case against Geological Disposal of Nuclear Waste*. Berkeley: University of California Press.

Silbey, D. J. 2007. *A War of Frontier and Empire: The Philippine American War*, 1899–1902. New York: Hill and Wang.

Silverio, I. A. 2011. "Killed for Anti-Mining Stand? Palawans Doc Gerry Was Friend of Environment and the Poor." http://bulatlat.com/main/2011/01/25/killed-for-antimining-stand-palawan's-'doc-gerry'-wasfriend-of-environment-and-the-poor (accessed 27 May 2011).

Sison, J. M. 2007. *Antirevisionist Struggle and Cultural Revolution: Consequences to the Communist Party of the Philippines*. Utrecht, Netherlands: National Democratic Front of the Philippines.

Skinner, B. J. 1976. *Earth Resources*, 2nd ed. Englewood Cliffs, New Jersey: Prentice Hall.

Sluyter, A., A. D. Augustine, M. C. Bitton, T. J Sullivan and F. Wang. 2006. "The Recent Intellectual Structure of Geography." *Geographical Review* 96, no. 4: 594–608.

Smith, A. [1776] 1991. *The Wealth of Nations*. New York: Alfred A. Knopf.

Smith, K. S. 2005. "Acid Rock Drainage." In *Mining in New Mexico: the Environment, Water, Economics, and Sustainable Development*, ed. L. G. Price, D. Bland, V. T. McLemore, and J. M. Barker, 59–63. Socorro, New Mexico: New Mexico Bureau of Geology and Mineral Resources.

Snell, M. B. 2004. "The Cost of Doing Business." *Sierra Magazine* May/June 2004: 34–79.

Soussan, J. 1988. *Primary Resources and Energy in the Third World*. London: Routledge.

Spooner, M. H. 1994. *Soldiers in a Narrow Land: the Pinochet Regime in Chile*. Berkeley: University of California Press.

Stark, J., J. Li and K. Terasawa. 2006. *Environmental Safeguards and Community Benefits in Mining: Recent Lessons form the Philippines.* Falls Church, Virginia: Foundation for Environmental Security and Sustainability.

Stavenhagen, R. 2003. *Report of the Special Rapporteur on the Situation of Human Rights and Fundamental Freedoms of Indigenous People.* New York: United Nations.

Stenson, J. 2006. "Disaster Management as a Tool for Sustainable Development: A Case Study of Cyanide Leaching in the Gold Mining Industry." *Journal of Cleaner Production* 14, nos. 3–4: 230–3.

Stevens, P. 2003. "Resource Impact: A Curse or a Blessing? A Literature Survey." *Journal of Energy Literature* 9, no. 1: 1–15.

Stiller, D. 2000. *Wounding the West: Montana, Mining and the Environment.* Lincoln: University of Nebraska Press.

Stinjns, J. P. 2005. "Natural Resources Abundance and Economic Growth Revisited." *Resources Policy* 30, no. 2: 107–30.

Strongman, J. 1994. *Strategies to Attract New Investment for African Mining.* Washington DC: World Bank.

Swibold, D. L. 2006. *Copper Chorus: Mining, Politics, and the Montana Press, 1859–1959.* Helena: Montana Historical Society Press.

Swyngedouw, E. 2004. "Globalisation or Glocalisation? Networks, Territories and Rescaling." *Cambridge Review of International Affairs* 17, no. 1: 25–48.

Tabunda, A. M. L. 2007. "The Poverty Scorecard." In *Understanding Poverty: The Poor Talk About What it Means to be Poor,* ed. P. P. Sicam, 16–48. Manila: Anvil Publishing.

Tan, S. K. 2002. *The Filipino-American War, 1899–1913.* Quezon City: University of the Philippines Press.

Tauli-Corpuz, V. and E. R. Alcantara. 2004. *Engaging the UN Special Rapporteur on indigenous people: Opportunities and challenges.* Baguio: Tebtebba Foundation.

Thomalla, F., T. Downing, E. Spanger-Siegried, G. Han and J. Rockstrom. 2006. "Reducing Hazard Vulnerability: Towards a Common Approach between Disaster Risk Reduction and Climate Adaptation." *Disasters* 30, no. 1: 39–48.

Thomas, W. L., ed. 1956. *Man's Role in Changing The Face of the Earth.* Chicago: University of Chicago Press.

Thompson, M. R. 1996. *The Anti-Marcos Struggle: Personalistic Rule and Democratic Transition in the Philippines.* Quezon City: New Day Publishers.

Tilton, J. E. and G. Lagos. 2007. "Assessing the Long-Run Availability of Copper." *Resources Policy* 32, no. 1: 19–23.

Torion, A. R. 2010. "PAG-ASA expects El Niño to end soon." MindaNews. http://mindanews.com/main/2010/04/24/pag-asa-expects-el-nino-to-end-soon/ (accessed 28 June 2010).

Transparency International. 2011. "Corruption Perceptions Index 2010." http://www.transparency.org/policy_research/surveys_indices/cpi/2010 (accessed 7 August 2011).

Tujan, A. A. 2001. "Corporate Imperialism in the Philippines," In *Moving Mountains: Communities Confront Mining and Globalization,* ed. G. Evans, J. Goodman and N. Lansbury, 147–94. Sydney: Otford Press.

Tujan, A. A., ed. 2007. *Jobs and Justice: Globalization, Labor Rights and Workers' Resistance.* Quezon City: IBON Books.

Tujan, A. A. and R. B. Guzman. 2002. *Globalizing Philippine Mining.* Manila: IBON Books.

Tyner, J. A. 2009. *The Philippines: Mobilities, Identities, Globalization.* London: Routledge.

Umitsu, M., C. Tanavud and B. Patanakanog. 2007. "Effects of landforms on tsunami flow in the plains of Banda Aceh, Indonesia, and Nam Khem, Thailand." *Marine Geology* 242, no. 1: 141–53.

United Nations Development Programme. 2003. *Philippines: Case Study on Human Development Progress Towards the MDG at the Sub-National Level.* New York: United Nations.

———. 2011. "Philippines, Country Profile of Human Development Indicators." http://hdrstats.undp.org/en/countries/profiles/PHL.html (accessed 5 May 2011).

United States Army and United States Marine Corps, eds. 2007. *The United States Army/Marine Corps Counterinsurgency Field Manual.* Chicago: University of Chicago Press.

United States Bureau of Mines. 1965. *Minerals Yearbook.* Reston, Virginia: United States Bureau of Mines.

———. 1970. *Minerals Yearbook.* Reston, Virginia: United States Bureau of Mines.

———. 1971. *Minerals Yearbook.* Reston, Virginia: United States Bureau of Mines.

———. 1977. *Minerals Yearbook.* Reston, Virginia: United States Bureau of Mines.

———. 1981. *Minerals Yearbook.* Reston, Virginia: United States Bureau of Mines.

———. 1984. *Minerals Yearbook.* Reston, Virginia: United States Bureau of Mines.

———. 1990. *Minerals Yearbook.* Reston, Virginia: United States Bureau of Mines.

———. 1991. *Minerals Yearbook.* Reston, Virginia: United States Bureau of Mines.

———. 1992. *Minerals Yearbook.* Reston, Virginia: United States Bureau of Mines.

———. 1993. *Minerals Yearbook.* Reston, Virginia: United States Bureau of Mines.

United States Census Bureau. 2011. "International Data Base." http://www.census.gov/ipc/www/idb/country.php (accessed: 9 May 2011).

United States Department of State. 2011. "Country Reports on Human Rights Practices – Philippines, 8 April 2011." Available at: http://www.unhcr.org/refworld/docid/4da56d989c.html (accessed 9 May 2011).

United States Environmental Protection Agency. 2004. *Evaluation Report: Nationwide Identification of Hardrock Mining Sites.* Report No. 2004-P-00005. Washington: US Environmental Protection Agency.

———. 2009. *2009 Edition of the Drinking Water Standards and Health Advisories.* Washington: US Environmental Protection Agency.

United States Geological Survey. 1996. *Minerals Yearbook.* Reston, Virginia: United States Geological Survey.

———. 1997. *Minerals Yearbook.* Reston, Virginia: United States Geological Survey.

———. 2000. *Minerals Yearbook.* Reston, Virginia: United States Geological Survey.

———. 2002. *Minerals Yearbook.* Reston, Virginia: United States Geological Survey.

———. 2011. "Historical Statistics for Mineral and Material Commodities in the United States." http://minerals.usgs.gov/ds/2005/140/#data (accessed 13 May 2011).

Utting, P. 2000a. "Towards Participatory Conservation: An Introduction." In *Forest Policy and Politics in the Philippines: The Dynamics of Participatory Conservation,* ed. P. Utting, 1–11. Quezon City: Ateneo de Manila University Press.

———. 2000b. "An Overview of the Potential and Pitfalls of Participatory Conservation." In *Forest Policy and Politics in the Philippines: The Dynamics of Participatory Conservation,* ed. P. Utting, 171–215. Quezon City: Ateneo de Manila University Press.

Valerio, R. 2005. Supervising Geologist, Davao Office, Mines and Geosciences Bureau. Personal interview, Davao City, Philippines, 3 June.

Van Aalst, M. K. 2006. "The Impacts of Climate Change on the Risk of Natural Disasters." *Disasters* 30, no. 1: 5–18.

Vanden, H. E. 2006. "Brazil's Landless Hold Their Ground." In *Dispatches from Latin America: On the Frontlines against Neoliberalism*, ed. V. Prashad and T. Ballve, 283–96. Cambridge, Massachusetts: South End Press.

Vanden, H. E. and G. Prevost. 2006. "The Political Economy of Latin America." In *Politics of Latin America: The Power Game*, ed. H. E. Vanden and G. Prevost, 145–76. 2nd ed. New York: Oxford University Press.

Veiga, M. M., M. Scoble and M. L. McAllister. 2001. "Mining with Communities." *Natural Resources Forum* 25, no. 3: 191–202.

Vesilind, P. J. 2002. "Hotspot: The Philippines." *National Geographic* 202, no. 1: 66–81.

Vick, S., G. Atkinson and C. Wilmot. 1985. "Risk analysis for seismic design of tailings dams." *Journal of Geotechnical Engineering* 111, no 7: 916–33.

Vidal, A. T. 2004. *Conflicting Laws, Overlapping Claims: The Politics of Indigenous Peoples' Land Rights in Mindanao*. Davao City: Alternate Forum for Research in Mindanao.

_____. 2005. *Resource Kit on Mining Issues in Mindanao*. Davao City: Alternate Forum for Research in Mindanao.

Virola, M. 2010. "Ex-Provincial Official in Oriental Mindoro Shot Dead." http://newsinfo.inquirer.net/breakingnews/regions/view/20100517-270550/Anti-mining-activist-shot-dead-in-Oriental-Mindoro (accessed 27 May 2010).

Vitug, M. D. 2000. "Forest Policy and National Politics." In *Forest Policy and Politics in the Philippines: The Dynamics of Participatory Conservation*, ed. P. Utting, 11–39. Quezon City: Ateneo de Manila University Press.

Vivoda, V. 2008. "Assessment of the Governance Performance of the Regulatory Regime Governing Foreign Mining Investment in the Philippines." *Minerals & Energy* 23, no. 3: 127–43.

_____. 1960. *The Constitution of Liberty*. Chicago: University of Chicago Press.

Wang, Y. H., I. H. Lee and D. P. Wang. 2005. "Typhoon Induced Extreme Coastal Surge: A Case Study at Northeast Taiwan in 1994." *Journal of Coastal Research* 21, no. 3: 548–52.

Ward, K. and K. England. 2007. "Introduction: Reading Neoliberalization." In *Neoliberalization: States, Networks, Peoples*, ed. K. Ward and K. England, 1–22. Oxford: Blackwell.

Warf, B. 1993. "Postmodernism and the Localities Debate: Ontological Questions and Epistemological Implications." *Tijdschrift voor Economische en Sociale Geografie* 84, no. 3: 162–68.

Warhurst, A. 1992. "Environmental Management in Mining and Mineral Processing in Developing Countries." *Natural Resources Forum* 16, no. 1: 39–48.

_____. 1994. *Environmental Degradation from Mining and Mineral Processing in Developing Countries: Corporate Responsibilities and National Policies*. Paris: Organisation for Economic Co-operation and Development.

_____. 1999. "Environmental Regulation, Innovation, and Sustainable Development." In *Mining and the Environment: Case Studies from the Americas*, ed. A. Warhurst, 15–49. Ottawa: International Development Center.

Warner, J. and M. T. Ore. 2006. "El Niño Platforms: Participatory Disaster Response in Peru." *Disasters* 30, no. 1: 102–17.

Weber-Fahr, M. 2002. *Treasure or Trouble? Mining in Developing Countries*. Washington DC: International Finance Corporation.

Westin, R. 1992. "Intergenerational Equity and Third World Mining." *University of Pennsylvania Journal of International Business Law* 13, no. 1: 181–223.

Whitmore, A. 2006. "The emperor's new clothes: Sustainable mining?" *Journal of Cleaner Production* 14, no. 3–4: 309–14.

Wisner, B., P. Blaikie, T. Cannon and I. Davis. 2004. *At Risk: Natural Hazards, People's Vulnerability and Disasters*, 2nd ed. London: Routledge.

Wolf, E. 1972. "Ownership and Political Ecology." *Anthropological Quarterly* 45, no. 2: 201–5.

Wood, C. and J. Bailey. 1994. "Predominance and Independence in Environmental Impact Assessment: The Western Australian Model." *Environmental Impact Assessment Review* 14.1: 37–59.

World Bank. 1992. *Strategy for African Mining*. World Bank Technical Paper No. 181 Africa Technical Department Series, Mining Unit, Industry and Energy Division. Washington DC: World Bank.

————. 2009. *World Development Indicators*. Washington DC: World Bank.

————. 2010. *Philippines: Fostering More Inclusive Growth*. Washington DC: World Bank.

World Commission on Environment and Development. 1987. *Our Common Future*. Oxford University Press.

World Resources Institute. 2003. *Mining and Critical Ecosystems: Mapping the Risks*. Washington DC: World Resources Institute.

World Wildlife Fund and International Union for the Conservation of Nature. 1999. *Metals from the Forest: Mining and Forest Degradation*. Gland, Switzerland: World Wildlife Fund.

Xstrata. 2008. *Tampakan Project: Sustainability Report 2008*. Zug, Switzerland: Xstrata Copper.

Yates, E. E. 1993. "Public Participation in Economic and Environmental Planning: A Case Study of the Philippines." *Denver Journal of International Law and Policy* 22, no. 1: 107–20.

Yocogan-Diano, V. 2009. "Introduction." In *Mining and Women in Asia: Experiences of Women Protecting Their Communities and Human Rights against Corporate Mining*, ed. V. Yocogan-Diano, 2–8. Chiang Mai, Thailand: Asia Pacific Forum on Women, Law and Development.

Young, J. E. 2000. *Gold: At What Price?* Washington DC: Mineral Policy Center.

Yu, J. M. 2005. Executive Director, Kalumonan Development Center. Personal interview, San Isidro, Philippines, 21 May .

Yumul, G. P., N. A. Cruz, N. T. Servando and C. B. Dimalanta. 2001. "Extreme Weather Events and Related Disasters in the Philippines, 2004–08: A Sign of What Climate Change Will Mean?" *Disasters* 35, no. 2: 362–82.

INDEX